SLOW GETTING UP

© AP Photos/Jack Dempsey

SLOW
GETTING
UP

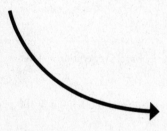

A Story of NFL Survival from the
Bottom of the Pile

NATE
JACKSON

HARPER PERENNIAL

NEW YORK • LONDON • TORONTO • SYDNEY • NEW DELHI • AUCKLAND

HARPER ⬤ PERENNIAL

FIRST HARPER PERENNIAL PAPERBACK EDITION PUBLISHED 2014.

Designed by Michael Correy

The Library of Congress has catalogued the hardcover edition as follows:

Jackson, Nate.
   Slow Getting Up : a story of NFL survival from the bottom of the pile / Nate Jackson.
     pages cm
   ISBN 978-0-06-210802-9
   1. Jackson, Nate. 2. Football players--United States--Biography. I. Title.
   GV939.M29J33 2013
   796.332092--dc23
   [B]

                                      2013011427

ISBN 978-0-06-210803-6 (pbk.)

14 15 16 17 18 OV/RRD 10 9 8 7 6 5 4 3 2 1

To Mom and Dad

Contents

Why did football bring me so to life? I can't say precisely. Part of it was my feeling that football was an island of directness in a world of circumspection. In football a man was asked to do a difficult and brutal job, and he either did it or got out. There was nothing rhetorical or vague about it; I chose to believe that it was not unlike the jobs which all men, in some sunnier past, had been called upon to do. It smacked of something old, something traditional, something unclouded by legerdemain and subterfuge. It had that kind of power over me, drawing me back with the force of something known, scarcely remembered, elusive as integrity—perhaps it was no more than the force of a forgotten childhood. Whatever it was, I gave myself up utterly. The recompense I gained was the feeling of being alive.

—FREDERICK EXLEY

Football is fun.

—JAKE PLUMMER

SLOW GETTING UP

Prologue

Goodbye, Dude
(2008)

Blank screen.

I hear the sound of children playing.

And I can feel the wind. I hear it rustling through the autumn leaves.

I smell the wet dirt and the long grass.

There is an iron taste in my mouth. This is where I belong.

Mom, tell me when it's time—

—Don't move, Nate. They're going to a TV timeout. Just relax.

That's Greek, our trainer. It's Thursday night in November 2008 and we're playing a nationally televised game in Cleveland. I'm a tight end for the Denver Broncos. I'm sure that, above me, the hit is being replayed over and again in slow motion. I also know that my mother is watching at home. With Greek holding my head and neck still, I move my legs and arms to let her know I'm not paralyzed.

After a minute, I get up and walk off the field, mad at myself for not holding on to the ball. I almost caught it. Had it in my hands. But Willie McGinest, a linebacker for the Browns, dislodged it when he buried his shoulder into my temple and spun me around in the air. I hit the ground like a dead body.

I stand on the sidelines as Jay Cutler finishes the drive with his third touchdown pass of the quarter. It goes to Brandon Marshall. After the score B-Marsh reaches for something in his pants but Brandon Stokley, another star receiver, stops him, fearing a flag for an unlicensed prop. The Browns receive the kickoff, can't score, and we win. A much-needed win; we had dropped the previous three. The locker room afterward is raucous with reenactments of the end zone shenanigans. B-Marsh had been reaching for a home-made black and white unity glove he had tucked into his game pants, and now, in the safety of the locker room, Stokley's stand-ing on a bench doing his best Tommy Smith impression from the 1968 Olympics. It is two days after Barack Obama's election and B-Marsh wanted to honor the moment. His president is black and he is proud. And like many proud black men who came before him, he got bear-hugged by whitey. Great gesture, bad timing. They call it the No Fun League for a reason.

On the airplane ride back to Denver I sit completely still and sip a cocktail. We used to have beers on the flights but NFL Commis-sioner Roger Goodell banned them. Legislate all you want, demand finds its supply. And booze is easier to smuggle past a tarmac TSA screening than a thirty-pack.

I go to our team physician, Dr. Geraghty, and ask him if he could give me something for the pain.

—I can't move my neck, Doc.

He says the best he can do is one Vicodin and one muscle relax-er and hands me two pills in a small bag.

—That's it? Two pills?

I hold up the nearly empty bag.

—You're going to make me hit the streets for this one?

—Sorry, Nate.

The two pills don't make it off the plane. I lie in bed all week-end, unable to move my head. By the time Monday comes around I put on my sweats and drive into work, stiffer than a wedding night's dick, as one of my coaches used to say. Business as usual.

Yes, I could have gone in for injury treatment over the weekend, but I'm sick of being treated for injuries, sick of spending time in the training room, sick of feeling fragile. It is my sixth year in the league. I'm well versed in the injury/rehab cycle of professional football. I know which injuries I need to treat and which ones I can handle on my own. This one I can handle. As long as I can run fast I'm fine.

I deal with the pain all week and by game day I am ready to play. It will be the last game of my career.

It is in Atlanta against the Falcons. We win 24–20. I have three catches for 33 yards. I jump over a cornerback after one of them. Most cornerbacks tackle low. They shoot for the kneecaps or the ankles because that's how you can bring down a larger man. The announcer says I shouldn't have jumped over him. He says it was too dangerous. I could have been hurt. Worse, I could have fumbled.

Several days after our win in Atlanta we're practicing in preparation for the Raiders game at home in Denver. Practice is dragging along. We're running plays against our scout team defense. There are two tight ends in the huddle: me and Tony Scheffler. Our quarterback, Jay Cutler, calls a play that has us running mirrored corner routes on either side of the ball. Tony and I are always being scolded for not reaching our required depth on our routes. If the route calls for ten yards, we're always breaking it off at nine. If it calls for twelve, we make it eleven. We're the same that way: eager to get there and eager to get the ball. We break the huddle and agree to go for the full twelve this time.

I run it full and break to the corner. Jay throws me a fastball with an arc that leads me to the sideline. I burst to track it down and a lightning bolt strikes me from behind. My hamstring rips off the ass bone with a bang, the sound of my season ending right there.

A month later, after a three-game losing streak puts us out of the playoffs for the third consecutive year, our team's season ends, too. And a few days after that, our head coach, Mike Shanahan—the man who brought me to Denver in the first place—is fired. There is zero job security in the NFL. Everyone knows that. But if there *was* anything close to job security, everyone thought Coach Shanahan had it. He had won two Super Bowls for the city of Denver. He was a close friend of the owner. He was building a new house. We were good every year. But good isn't good enough in the National Football League.

After a two-week search, the Broncos hire Josh McDaniels, thirty-two-year-old offensive coordinator with the New England Patriots. He helped organize the most potent offense in the NFL while still in his twenties. A young Shanahan, some are saying. I'm still rehabbing at the facility every day so a few days after he is hired, I go upstairs to introduce myself. Check him out. See if he knows I'm on the team. We're pretty much the same age. We both went to Division III colleges. I'm sure we have a lot in common.

The Denver Broncos facility is an ode to the Super Bowl victories of 1997 and '98. Life-sized pictures of John Elway, Ed McCaffrey, Rod Smith, Shannon Sharpe, and Terrell Davis line the halls of the second floor, where the coaches' offices are. The year before Shanahan was hired as the head coach of the Broncos, he was the offensive coordinator of the San Francisco 49ers, whose 1994 team was one of the most dominant in the history of the sport. It catapulted Coach to superstardom. He brought that 49er pedigree to Denver and changed the sports landscape of the city.

Just outside of his office there was an eight-foot photo of himself. It was the classic, intense, game-day countenance of the wizard behind the curtain. When you stood at the threshold, this face was a reminder of what was expected: accountability through tradition. Every time I had stepped foot in that office, it had been to sit down at Mike Shanahan's desk.

But now he is gone, and so are all of his assistant coaches. The pictures have been taken off the walls. The halls are empty. Except for our general manager, Brian Xanders, who has somehow kept his job. And sitting behind the desk usually occupied by Shanahan's longtime secretary is a bro wearing a baseball cap and sucking on a red lollipop.

His cap is severely bent, his striped polo shirt is fading, and the pockets of his cargo shorts bulge with what can only be more candy. Waxy wrappers are strewn on his desk. The NFL is supposed to be an institution of proud tradition, I say to myself, where battle-tested men carry themselves with class and dignity.

—An institution of what?

—Dude, do you have any Smarties?

I am sitting in a chair in the hallway, dumbstruck at the change in fortune. I imagine young Josh McDaniels walking into Pat Bowlen's office for the interview and tossing a jockstrap on the owner's desk without saying a word. After a minute of awkward silence, he probably said something like, Go ahead. Smell it. This belonged to Tom Brady, Mr. Bowlen. That's the smell of success. That's the smell of the Super Bowl. That's the smell of . . . *balls*, Pat.

Done. The job was his.

—Josh will see you now.

I step into Shanahan's old office and shake hands with my new coach. He looks like a little kid sitting in the cockpit of an airplane. For fifteen minutes he rattles off clichés about his football philosophy and his plans for the team. He says that everyone will have a chance to prove themselves on the field. He says he looks forward to seeing how I will digest his offensive system. But he doesn't make eye contact with me and doesn't laugh at my jokes. After an awkward silence, I grab my jockstrap off his desk and leave. I know I'm in trouble.

A few weeks later my rehab is almost complete. My ass bone has tentatively embraced the return of the tendons on my hamstring; a deal brokered by the bridge of goop injected into the bony insertion of my ischial tuberosity. The procedure is called a PRP injection: platelet-rich plasma. It's my own blood, spun in a centrifuge to separate the good goop from the bad, then injected into the site of the injury: the insertion of hamstring to lower pelvis. It's the second time in two seasons I've had a PRP shot.

Before I can go home, I have to be medically cleared. I submit to several strength and agility tests and Greek says that I look fine. All I have to do is sign an affirmation of health. For physically sound players this is done the day after the last game of the season. Any injury that's in the books—meaning anything that received treatment in the training room—needs to be considered healthy in order for you to go home.

—How's your left ankle?

—Good.

—Your right wrist?

—Good.

—Your neck?

—Good.

—Your ninth and tenth ribs on the left side?

—Good.

—Your fifth and sixth on the right?

—Good.

—Do you consider yourself fit to play football?

—Yes.

—Sign here.

It's one more piece of paper in an ever-growing injury file about the size of a dictionary. I sit down across from Greek and Dr. Boublik at the end of every season. It's the only day of the year I see that file. By now, the thickness disturbs me. I wince a little and scribble my name on the dotted line. The dictionary slams shut. I'm free to go.

Early the next month I'm in San Jose, California, visiting my family. I'm leaving my sister's house and my phone rings. It's my agent, Ryan Tollner.

—Hello?

—Hi, Nate. Have you heard anything from the Broncos?

—No.

—They're releasing you today. No one called you?

—No.

—Wow. All right, we're going to find you another team.

I hang up and drive. My Bronco life, like a stack of images being shuffled, flashes in front of my windshield. I always knew this moment would come but there was no way to prepare for it. In the NFL, you are alive until you are dead. There is no in between, and no way to put yourself on the other side mentally. You fight every day to keep your job by convincing yourself that you belong. And every day you return to work and see your name still posted above your locker is proof that you deserve that locker. Then one day, fate sneaks up behind you, taps you on the shoulder, and breaks your nose—or blows out your knee.

Then it's over.

When I get back to my parents' house, my mom tells me there is a message for me on the machine.

—Uh, hi Nate, this is Brian Xanders with the Denver Broncos. Please give me a call when you can regarding your, uh, garble garble garble. My number is bleep bleep bleeble.

I call him and get his machine, leave a message, and wait. I sit on the bed in my childhood bedroom, looking out on the street where I fell in love with the sport. Out there I was Jerry Rice, streaking down the gutter line in the fading daylight. In here I am a twenty-nine-year-old man, waiting for my NFL coffin to be nailed shut.

Late into the afternoon: still no call from Xanders. I call him again; again his machine. Busy guy. After ten minutes, he calls me back.

—Hey, Nate. So you probably know by now. We're releasing you today. It wasn't anything you did, necessarily. We've gone over every player in detail and we've decided to go in another direction. I want to thank you for all your hard work and if there's anything I can do to help you from here on out, let me know. I'll have nothing but good things to say if we are contacted by another team. Oh, and if you want a better explanation, feel free to call Josh. He'd be happy to talk to you. Okay. Good luck, Nate.

Going in another direction. Bang! The final nail was a cliché. I call Josh. *C'mon, Josh, tell me I'm too old. Tell me I'm too slow. Tell me I'm damaged goods. Tell me I'm not good enough. Tell me something. Just don't bullshit me.*

—Coach McDaniels's office.

The Dude.

—May I speak to Josh, please?

—Who can I tell him is calling?

—Nate Jackson.

—Okay, just one second.

Thirty seconds later.

—Nate, yeah, he's, uh, he's in a meeting. But he said he'll call you right back. What's the best number to reach you at?

—The same number you have on file. My cell: 867-5309.

—Okay, got it. Thanks.

He never called.

1

The First Seven Years
(2002)

"Nate Jackson, wide receiver, Menlo College."

I walk to the front of the stage and stop as instructed. I'm in my underwear. A trickle of sweat runs down my side. One hundred men sit in folding chairs with clipboards in their laps. They look me up and down and scribble notes. Who are these sick fucks?

Out in the hall I put my clothes back on and walk past the line of dudes in their underwear who haven't gone in front of the audience yet. It's January 2002. I'm at the Hyatt in downtown San Francisco for the East-West Shrine Game, a college all-star game for seniors. It's a football game, yes, but it doubles as a weeklong job fair for all thirty-two NFL teams. After the Hanes runway fashion show I'm ushered into the room where the New York Giants administer their four-hundred-question personality test. Because when it's crunch time and the game is on the line, the front office needs to know one thing: do I prefer Jell-O or pudding? I look around the room during the test and decide that most of us prefer both.

· · ·

Outside in the lobby I find my new agent, Ryan Tollner, sitting on a couch with a defensive end named Akbar and someone else and they're arguing about who the most famous athlete in the world is: Muhammad Ali or Michael Jordan.

Ryan's in his midtwenties. His light brown hair is nicely combed and his sharp jawline is clean-shaven. A former backup quarterback at Cal, Berkeley, he speaks with a deliberate clarity that puts me at ease. He's just getting into the agent business with his cousin Bruce Tollner, after working a year as a financial analyst. He doesn't have many clients yet. And I don't know any other agents. In other words, it's a good fit for both of us. A month ago, after my college season was over, Ryan took me to a 49ers game so we could get to know each other. I hadn't been to a Niners game since they beat the Packers in the playoffs in early 1999, when I was nineteen years old.

My friends and I had pooled funds and bought tickets at the last minute. We drank beer in the parking lot and pulled out a poster board and Sharpies. After much deliberation, we decided: "Wisconsin, please . . . this is the REAL California cheese." It didn't make much sense. There was a drawing of a rat sniffing at a triangle of stinky Swiss. Once we got to our upper-deck seats no one wanted to hold it up. It lay folded on the concrete late into the game as the energy in the stadium became desperate. The 49ers were trailing deep into the fourth quarter. The consensus was that all was lost. Terrell Owens, who had dropped some big passes, was a bum. Head coach Steve Mariucci was a loser. Steve Young was washed up. The Packers were going to beat us again. Then Steve Young hit Terrell Owens for the winning touchdown and capped Candlestick Park in a bubble of ecstasy. Tears of joy filled the stadium. I wiped them from my face. Wisconsin, please, this is the *real* California cheese.

Much had changed for me between the two Niners games. Sitting in the stands with Ryan, my future agent, I no longer felt like a fan. I no longer cheered. I was watching my potential peers. I wanted in. After the game we stood outside of the 49ers locker room and watched the players walk out in street clothes. Ryan spoke with some of them. I stood off to the side. Then we left. Later that week I signed the necessary paperwork and officially made Ryan my agent. He will earn 3 percent of my contract for the rest of my career.

Now it's just Ryan and me in the lobby of the Hyatt. After the Ali/Jordan argument ends, I tell him about my hamstring, which I had pulled two days before checking into the Hyatt but hadn't told anyone about. My Menlo coach Dave Muir had been throwing me routes so I could stay crisp for the important week ahead. On one of the last routes of the day—after Dave had run off to the port-o-potty clenching his butt cheeks, then returned five minutes later apologizing—I pulled my hamstring as I came out of a break. It was a new sensation. I had never injured a hamstring before. I decided to keep it to myself and try to work through it. I was already a long shot to be in the Shrine Game, let alone the NFL. I didn't want anyone thinking I wasn't ready.

Ryan tells me to be smart with it. No use making it worse. But I have scouts to impress on the practice field. The week of practice is more important than the game. The NFL guys are watching us like they're examining racehorses in the paddock.

On the second day of practice, I run a slant during one-on-ones and rip my hamstring for real. I'm off the field for the rest of the week, including the game. I'm sure my chance has just slipped away but Ryan tells me not to worry. Just rehab it and we'll get you on a team. There are other workouts, pro days and combines.

But Menlo, my alma mater, doesn't have a senior pro day, the day that the NFL scouts arrive to work guys out. Menlo's too small—about five hundred students—and it's a Division III program. I transferred there three years earlier after getting cut from Division 1AA Cal Poly. Poly's head coach, Larry Welsh, told me I was too slow to play receiver and too small to play tight end. Now scram, punk. Menlo revived a football dream that was dead in the water. I wanted to play: that was it. I didn't care about anything else. My high school coach, Myron Zaccheo, knew I was depressed about getting cut from Cal Poly. He was the one who told me about Menlo. It was thirty minutes away from my house in San Jose but I had never heard of it. But the more I found out, the more I felt pulled to Atherton.

Fred Guidici, Menlo's assistant head coach and recruiting ace, was my point man on the phone. With the voice of a long-lost friend, he told me about the pass-heavy offense and the local 49er pedigree. He told me about the stable of former NFLers coaching at Menlo. Ken Margerum, former Stanford and Chicago Bears receiver, was the head coach. Doug Cosbie, former Pro Bowl tight end for the Cowboys, was our offensive coordinator. Doug coached under Bill Walsh at Stanford. *The* Bill Walsh, the legendary head coach of the San Francisco 49ers who revolutionized offensive football. All-Pro defensive end for the Vikings Keith Millard was our defensive coordinator. Former All-Pro 49er guard Guy McIntyre was our offensive line assistant. The list went on. I visited the campus and instantly felt at home. I decided to transfer, leaving behind an ideal California college setting and a group of bewildered friends. "You're doing what?" "You're going where?" "Men-huh?"

But Menlo was an oasis. The campus was lush and quiet. The classes were small and inclusive. The professors were friendly and passionate. The surrounding neighborhoods are wealthy and extremely safe. A common police blotter blurb from the local newspaper: "Suspicious looking cat seen near parked car." Or: "Man tying his shoes across the street from suspicious looking cat."

After my first season at Menlo, Ken Margerum went on to coach receivers at Cal Berkeley and Doug Cosbie took over as head coach. Dave Muir, a backup quarterback at Washington State who was only a few years older than me, joined the coaching staff as our receiver's coach. We became friends almost overnight, a rarity between players and coaches. A few days into two-a-days he called me up to the coaches' offices.

"Nate," he said, "you can play on Sundays. I've seen it and I know it. We're going to make it happen." The next two seasons were a dream. I caught over 100 balls in consecutive seasons and earned All-American honors alongside my quarterback, Zamir Amin, who ran Walsh's brilliant offense once to the tune of 731 passing yards in a single game. I was playing football with my best friends: for football's sake. I was as happy as I have ever been.

After a big game in week two of the 2001 season, Coach Cosbie, who stood six foot six, with light blue eyes and a booming voice, told me that someone wanted to talk to me. Bill Walsh came to our home games from time to time. His son Craig used to be Menlo's athletic director, and Bill still supported the school. I always knew when he was there. He stood in the northeast corner of the end zone. His silver hair was unmistakable. So was the way he stood, with his arms folded beneath an omniscient gaze. I had been watching him since I was a boy. I dropped my helmet on the grass and jogged over to the legend. He told me to keep doing what I was doing. He said that I would get my shot at the next level. I jogged away with a new pep in my step. Men-huh?

Four months later, as I stand on the sidelines watching one of the East-West practices that I can no longer participate in, Bill Walsh comes and stands next to me. He's one of the chairmen of the Shrine Game committee and helped get me on the roster. Like Ryan, he tells me not to worry too much about the hamstring.

—Just get healthy and you'll have your shot.

I crawl off the ledge and, after the game, start rehabbing immediately.

A couple of months later I'm at San Jose State's pro day. It's early in the morning, cold outside. But the presence of stern NFL men fingering their stopwatches puts a buzz in the air. That's the first thing that strikes me: The NFL is a pageant. Football is no longer just a game. The NFL's regional scouts stretch from Pacific to Atlantic and work all year long compiling information on their employer's potential investments.

They're conducting research for their bosses. I'm trying to touch God. This is the moment I've been training for. Most of us here weren't invited to the NFL Scouting Combine in Indianapolis. Only 260 guys go to that. This is our only shot in front of scouts. We run the 40-yard dash, the L-shuttle, the 20-yard shuttle, and the 60-yard shuttle. We also do the broad jump, the vertical jump, and hit the 225-pound bench press. Then we run routes.

The success I had at Menlo had come with an asterisk: *D3 football. I bought into it for a moment: just like I momentarily bought into Cal Poly coach Larry Welsh's assessment of my athletic ability. But after seeing the D1 talent at the Shrine Game, and now here at San Jose State's pro day, I'm hit with a surge of confidence. They're just dudes: dudes with strengths and dudes with weaknesses. Dudes with doubts and fears and pain. Humanity equalized all of us. D1, D3, NFL: just dudes.

Leading up to the draft the 49ers invite about thirty local prospects to their facility for a workout. I drive the fifteen minutes to Santa Clara and walk in the front door, greeted immediately by a trophy case that holds the talismans of my childhood. All five Lombardi trophies gleam in the halogen lights. Three of them came under Bill Walsh and two were under George Seifert, Bill's successor. We change into 49er shorts and jerseys in the locker room. I'm caught between focus for the workout and a surreal admiration of my surroundings. I walk through the weight room and imagine Joe Montana and Jerry Rice discussing a play between sets of leg presses, or Charles Haley yelling at Steve Young in the showers and having to be restrained by Ronnie Lott.

Out on the field it's all football stuff: ball drills, routes, one-on-ones. No stopwatches and no clipboards: just coaches coaching football. I have a good workout. There was no way I wouldn't. I feel ordained by the tradition of the institution. After I'm done, receiver coach George Stewart—a large, deep-voiced, kindhearted man everyone called Stew—pulls me aside and tells me what I'm start-ing to figure out. I can do this.

At the end of the month is the two-day draft. The day before it starts, the Niners want to time me in the 40-yard dash. It is a beautiful, sunny day. I drive from the small parking lot outside of my Menlo dorm room, down El Camino Real to Stanford Stadium, where I meet Ryan and a few 49ers scouts. I run a 4.5 and a 4.6 on the bright green field of an empty stadium. I ask Ryan how I looked. Strong, he says.

The first day of the draft passes. The next morning Ryan tells me that the Niners might draft me late. Some of my close Menlo friends come over to my house to watch it with me. Dave Muir sits next to me in the deafening silence as we watch player after player get what I want. I stare at the phone. During the seventh round it rings: the Ravens. They want to sign me as a free agent after the draft ends. I want to be drafted. If I'm not, I can still sign as a free agent. Sometimes, Ryan says, that's actually better, assuming you have a few teams that want you. Then you get to choose. If you are drafted, you have no choice.

The last pick comes and goes and my name isn't called. A few minutes later the phone rings again, the 49ers this time. They want me, too. My first business decision: Baltimore or San Francisco. Be-cause of my many connections to the team, the fact that the 49ers didn't draft any receivers (the Ravens drafted three), and the team's proximity to my family, Ryan and I agree that the Niners are my best bet. I call them back, while Ryan delivers Baltimore the bad news, and not a minute later I'm out the door, making my short way to the facility again. I walk upstairs to sign my new NFL contract, complete with a $5,000 signing bonus. Look, Ma, I'm a 49er!

Minicamps start the next week and my confidence continues to grow. I step to the line of scrimmage with an NFL cornerback in my face. He's coiled like a snake. My heart races. The ball is snapped and we strike. Sometimes he bites me. But sometimes I bite him. And each successful bite further confirms my suspicions: I belong here. One of my fellow receivers, Terrell Owens, is the most physically dominant wide receiver in the game. Early on I learn a valuable lesson from him, as we're of similar size. The lesson is this: Do what you're good at and do it well. Don't try to be something you're not. T.O. dominates the cornerbacks who try to cover him, throwing them around like rag dolls. He doesn't try to dance around on the line of scrimmage and look pretty. He picks them up and moves them out of the way. Then he runs his route. I'm not on that level, but I take note. Decide what you're going to do and do it violently.

I'm able to focus solely on the physical because I've got a solid jump start on the mental. Our offense at Menlo used the same terminology as the 49ers. Football-speak is a language. If you are not fluent, you are lost. And it varies from system to system. Some of the best athletes on the planet could never learn the NFL language, so they never got on the field.

Minicamps end in late June. We have a month-long break before training camp starts. I can hardly wait to get back on the field. I go into the facility and work out by myself. I shadowbox in the steam room. I stare into the mirror and flex my muscles. I'm ready.

Training camp is in Stockton, California, on the campus of the University of the Pacific. On the day we report, I drive my green 1996 Honda Civic to the campus and pull into the players' parking lot, sliding in between a Mercedes and a BMW. I check into the dorms and get my room assignment. I'm on the third floor in the rookie hall. My roommate is a center from Miami named Ty Wise. We are both free agents, both long shots. The fear of the unknown is almost crippling. It takes my breath away. I want to get the pads on and get on the field right away. Enough with all of the talking! But the NFL, it turns out, is mostly talking. I sit at attention while head coach Steve Mariucci (we call him Mooch) sets the scene for us in the auditorium-style classroom. Tomorrow we hit. Get ready.

On the morning of our first practice, I lie awake in my squeaky metal twin bed listening to Ty snore as the light creeps through the blinds. At six thirty I hear the sound that will soon attach itself to my football dreams. It's a sound that is linked to hope, to sweat, to pain, to glory. It's a sweet sound. It's the sound of the devil. At six thirty on the dot, a coffee-drunk assistant walks through the dorm halls and lets the air horn blow. Wake the fuck up. It's time to hit.

Practices are open to the public. I walk onto the field with my helmet in my hand and the stands erupt. What a welcome, I think. Bay Area kid makes good. Yeah, it's a great story. I glance to my left. There is T.O., and our starting quarterback, Jeff Garcia, walking beside me. They probably think the cheering is for them.

From day one I become aware of a powerful dynamic within each position group. We are all competing against each other for roster spots. We are sword fighting. There are five veteran receivers and five of us rookie free agents. Teams usually only keep five wide receivers on the roster. We all know this. And we know that in order to make the team, one of us has to unseat a proven NFL veteran. That means taking advantage of precious few opportunities. For those of us at the bottom of the depth chart, the reps are harder to come by. Some guys get discouraged. Others hang on to the coaching clichés. "Don't count your reps. Make your reps count." This is the ominous base note to the training camp death song. No one talks about it but we hear it loud and clear. Half of us will be gone by the end of the month.

Adding to the reverb of the death knell is my shitty left shoulder. I dislocated it twice in college and the Niner doctors spotted it during the pre-minicamp physical to which every player must submit.

IMPRESSION: This 22-year-old right-hand dominant running back [sic] from Menlo College has a history of two prior left-shoulder dislocations. He is currently asymptomatic. The patient was seen and examined with Dr. Dillingham. At this point he will be graded 4–5. Dr. Dillingham and head trainer Lindsy McLean will have him sign a waiver.

I had to sign the waiver to get on the field. Now if I hurt the shoulder again, they can cut me with no liability. A week into training camp I slip on the wet grass making an in-cut, a 90-degree break toward the middle of the field. I drop my hand to the ground to stop my fall and my shoulder pops out of the front of the socket. It feels, strangely, like my head is on backward. While I lie on the turf Lindsy McLean, the head trainer, arrives to slide it back in. Of the two dislocations I had in college, one slid back in easily, the other didn't. The opposing team's trainer had her foot on my chest and was pulling on my arm as if to free the sword in the stone. But she was no King Arthur. The muscles were spasming around the dislocated humeral head and wouldn't let go. So I rode in the back of an ambulance to the hospital, dressed in my full Menlo Oak regalia, where I was sedated, my muscles relaxed and my skeleton realigned. Thankfully there are no humeral complications this time. Lindsy reaches under my shoulder pads, lifts, twists, and pulls. Whoosh. I feel a powerful relief and jump to my feet. The more you dislocate your shoulder, the more likely you'll do it again. But the recovery time improves with each dislocation.

> Nate dislocated his left shoulder two days ago. He has near full range of motion. He had signed a waiver on this shoulder and had dislocated it in college on several occasions. He is going to rehab and get a shoulder immobilizer. He will be allowed to play but he does know that he has a risk of further dislocation and further damage both soft tissue and bony damage in the shoulder [sic]. . . . We did talk also about the possibility of stabilizing his shoulder if he desires at some point before or after the season depending on how things go.

A few days later I strap on a neoprene shoulder harness that looks like I'd got it at a sex shop in North Beach and go back on the field a day before our trip to Osaka, Japan. We are set to play the Washington Redskins in the American Bowl in our first preseason game. During minicamps we had our passport photos taken at the facility between practices and filled out the necessary forms. By training camp we have our brand-new passports. We board the huge plane and set sail to Glory. They stamp my passport as "Entertainer."

Five days in Japan flies by in a jet-lagged blur. Coach Mariucci gives us a good amount of time to ourselves, time we spend wandering the streets of Osaka, frightening the locals. Eminem blasts from storefront speakers. Adolescent Japanese girls sing along, unaware of what they are describing. We walk through the crowded streets as a pack of lanyard-clad Godzillas. The locals point and stare and run inside screaming. Big black man! Big black man! I'm given my own room at the hotel because another rookie did not make the trip. I push the small beds together and stretch out. I turn on the bidet, chuckle, use toilet paper instead. The Redskins are staying at our hotel. So are their cheerleaders. I spot one in the lobby who shoots an arrow through my heart. We fall in love immediately. She chooses not to acknowledge it, though, and so I give her the space she needs. I'm still waiting.

On one of our nights out I tag along with some of the veterans, a few girls who work for the Niners and some Niner cheerleaders. We go to a club on the top floor of a nondescript office building. The rookie-veteran barrier seems broken down, if only temporarily. We're united as strangers in a strange land. To the locals, we are interchangeable monsters. There are fifteen of us. When we walk in the record screeches to a stop and the entire club recoils in fear. The Japanese patrons slowly back into the corners of the room and watch us the rest of the night as if we are a multi-culti variety show. We party NFL-style: shots and dancing and yelling. And a few buff dudes with their shirts off. Jeff Garcia dances on the bar and passes out beers. Terrell Owens knows better than to get drunk during camp. But he takes off his shirt anyway.

We play the game a few days later. I watch from the sideline with my neoprene harness cinched up tight. Stew had told me I wasn't going to play. Rest the shoulder, he said. Don't worry about this game. You'll get your chance. But it's hard not to worry. I can feel my chance slipping away, like the Japanese girls when we walked into the room. *Konnichiwa!*

We fly back to California and back to training camp in Stockton and I am back in my metal bed, lying awake once again as the light cuts through the blinds. Soon the assistant coach with the air horn makes it official. Wake the fuck up. It's time to hit—again.

The next week we have a home game against the Kansas City Chiefs. I put on my wifebeater and go out to warm up early. I stand on the grass of Candlestick Park and feel fully immersed in my new profession. I look up to the spot where Ryan and I sat the previous year and smile to myself. I don't get in the game until late in the fourth quarter. On my first play, I catch a slant from Brandon Doman for five yards. On the next play I catch another slant for 10 yards. Football is easy. Our drive stalls out. The game goes into overtime. On our first possession of overtime, Brandon throws an interception down the middle. The DB is dancing around and cutting back against the grain, trying to end the game on a walk-off interception return for a touchdown. He cuts back one too many times. I stick him under his chin and body-slam him. They kick a field goal on the next play and win the game. The next day we watch film and Stew dissects our performances. Stew's favorite play from the game, besides T.O.'s freaky 71-yard touchdown, is my tackle in overtime. Receiver coaches love that shit.

On the last day of camp in Stockton, the offensive linemen go out for dinner. As is customary, they get the rookies shitfaced. Ty comes back to the room like a tornado and plops down on the bed. John Engelberger is a veteran defensive end—a big, corn-fed bruiser, who likes hanging with us rookies. We sit around the room laughing. Ty is a funny man, even more so after too many tequila shots. After one of his jokes, he hiccups and reaches for the closest receptacle. It's an empty water bottle. He attempts to vomit into it but instead creates a suction. It sprays out the sides of his mouth and all over the room.

—Ty! C'mon!

—Shut up, *Rapper*!

I had mistakenly told my teammates about my hip-hop aspirations, and now they tease me relentlessly. At least we can laugh. Training camp is over. That night I lie awake in bed and listen to the last of the partying linemen run wild underneath our window. They have found a golf cart and turned the campus into their own personal bumper car playground. I fall asleep a better man. The next day we pack up our dorm rooms and go back to the 49ers facility.

We're back to a regular schedule, done every day by five instead of ten. On one of our nights off I go to a movie with some friends and see Mooch outside the theater with his family. We talk for a few minutes. I meet his kids. He says see you tomorrow. The next day he approaches me as I'm standing on the sideline with ice on my shoulder.

—How's that shoulder?

—Ah, it's okay.

—You know, Nate. You've had a good camp. But you're hurt. I appreciate you toughing it out. And I'd love to keep you on the practice squad. But practice squad players need to be healthy, they need to be able to *practice*, you know what I mean?

—Yeah, I know what you mean, Coach.

A few days later I'm stopped as I walk through the locker room. An assistant tells me to come upstairs, and bring my playbook. My days as a 49er are over, it's obvious—only formalities remain. I saw this moment coming in slow motion ever since I slipped on the wet grass and my shoulder went pop. I am a broken machine. On the way out the door GM Terry Donahue tells me that if I get healthy they'll sign me back the next season. I know I have Bill to thank for that gesture. He is serving as a consultant for the Niners.

I go home and move back in with my parents. They have always been supportive of my football dreams, but we are not a football family. I'm the only athlete of all my siblings. My parents are school-teachers. We live in a middle-class California neighborhood in a small one-story home. I spent my summer days as a child at the cabana club down the street. A lifeguarded pool is a great babysitter. My friends and I were fish. I swam competitively and played soccer, but I had my eye on the oblong ball with laces. I was a 49ers child. But my parents wouldn't let me play competitive football until I was in high school. That I ended up in a 49ers uniform after not playing until high school, not getting recruited out of high school, and getting cut from Division IAA Cal Poly probably surprised them. So when I move back home after getting cut from my hometown team, they are extra-supportive. My dream came true, for a second. And now I'm licking my wounds in my childhood bedroom.

I have shoulder surgery paid for by my own insurance and I rehab at a clinic three times a week. My physical therapist is used to fraudulent worker's compensation cases and old people who have fallen down. She marvels at my recovery time and my dedication. I explain that I'm headed back to the NFL. I'm trying to convince myself that it's true but I have no idea. I'm holding on to Donahue's words. But maybe he was just being nice. Maybe my football days are over. My shoulder heals very fast but my mind is a mess. I sit around in my bedroom and have panic attacks. I try dating but can't relax. I scribble in my journal, trying to exorcise the demons, summon the angels, build future mental stairways. I watch the 49ers on TV all year with a new appreciation of the machine. For the first time I'm seeing the big picture through the small screen. I listen to the announcers and read the papers. The media narratives are sensational and simplistic, and when compared to what I know about the team, sound like drivel.

From the couch, my dad and I watch the football season unfold. The 49ers go 10-6 and slide in the back door of the playoffs, beating the Giants in the wild-card round at Candlestick Park in a thrilling comeback. But they lose badly the next week to the Tampa Bay

Buccaneers. A few days later Mooch is fired. I never understand firing a winning coach, but apparently there were some philosophical differences between Mooch and Donahue. There's so much more to it than anyone ever knows. After the dust settles on Mooch's firing, Ryan reaches out to Donahue and reminds him that I am still around. They need camp bodies. They always need camp bodies, especially at wide receiver. Receivers drop like flies during training camp. Donahue keeps good on his word and I drive back to the facility in my Civic. They put another contract in front of me. No signing bonus this time. Here's the pen. Look, Ma, I'm a 49er again.

The next week is the Super Bowl in San Diego between the Oakland Raiders and the Tampa Bay Buccaneers. Ryan and his cousin Bruce have just joined forces with super-agent Leigh Steinberg. Leigh throws a Super Bowl party every year. Ryan invites me. It's at the San Diego Zoo. I'll be on the list plus one, he says. I have friends who live in San Diego so I make the trip.

I bring my friend Justin to the party. He's in town working for Pepsi, driving around in a souped-up two-door Lexus with a custom Pepsi Blue paint job. He sets up his Pepsi table in the Gaslamp Quarter and passes out free "Pepsi Blue" samples from a carbonated backpack hose.

—Free Pepsi Blue! Like to try a sample?

—Sure.

—It tastes like fizzy Nyquil.

—Sorry.

Justin breaks free from his Pepsi duties and comes with me to the zoo. We check in at the will-call desk and I get my lanyard. It says I play for the San Francisco 49ers. I'm funneled into a line where I walk the red carpet like a plank, photographers looking at me confused but still snapping away. A face lifted man from *Entertainment Tonight* interviews me. We both wonder why. Inside the party I'm formally introduced to pro sports high society. It has a strange, seductive sheen. The women are beautiful. The men are powerful. Everyone is horny.

The next night we go to the Playboy party near Balboa Park with girls we had met at Leigh's zoo party. They told us to come along: that we'll find a way in. But we have no tickets and are not on the list. It's a gorgeous starry night. Attempts to squeeze past distracted security thugs are easily thwarted. The bouncers are on their game. This is their Super Bowl, too. We need actual tickets. Luckily I run into a girl whom I had met outside a club the previous night. There had been a rowdy mob trying to get in and she was pinned against a barricade. I pushed back against the mob and gave her some breathing room. Her name is Sasha.

Sasha has an extra ticket to the party, she says. It's for her sister but she's not going to make it. Here, take it. Now we need one ticket. I walk to the will-call desk.

—Hi, there, how are—

—Are you on the list?

—Yeah, should be.

—What's your last name?

—Jackson.

She flips through the stapled pages and runs her finger down a long, tight list. She stops her finger.

—Tom?

—Yep, that's me.

She hands me my ticket. Sorry, Tom. You've probably been to a hundred of these things, anyway. Inside the palace doors are the excesses of the industry; sports and entertainment collide in a puff of sex. Justin and I walk around and giggle. There are naked girls in body paint, celebrities, free drinks and free food and everyone laughing a little too loudly. I'll remember these girls forever. One of them ends up in a cab with us after the party. She has olive skin and dark hair. She speaks clearly. Smells of black orchids. Wears loose-fitting linen. Her earrings are dream-catchers. Her aura is magenta. We drive across the Coronado Bridge and drop her off at her resort hotel. I never see her again, but I smell her every time the wind whispers, *Mary.*

A few weeks later the Niners hire Dennis Erickson as Mooch's replacement. Although he had a few losing seasons, Mooch did his best to carry on the Niner tradition. He ran the same offense. He often referred to our forefathers. He kept counsel with the elders. He was Bay Area through and through. Erickson has none of that. He brings his own brand and the 49er brand blows away in the wind.

I keep reminding myself: Joe Montana was here. Jerry Rice. John Taylor. Brent Jones. Ronnie Lott. Roger Craig. Steve Young. Dwight Clark. Tom Rathman. Eddie D. Everyone was here. But I have to put my face right up against the glass trophy case to remember they ever existed in this building. Who are these people who call themselves 49ers? Not the 49ers I know.

Part of my disappointment with this new brand of football (a new system of offense, new terminology, new schedule, and new coaches) is that it's bumped me even further down the depth chart. I'm getting no reps. When Mooch was around, I often saw Bill Walsh on the sidelines at practice. He would offer me an encouraging word after a nice play, a nod or a pat on the back. But I rarely see him down there anymore. It's just me and my fellow long-shot receivers, blowing dandelions and chasing down the safety on backside run plays. As a receiver, it doesn't just matter if you get in the game; it matters what plays are called when you're in. The veterans get all of the good pass plays. When they get tired we go in for a few run plays or screen passes. Once they catch their breath we're back behind the huddle picking our butts.

On a typical training camp day in mid-August, I'm suited up, helmet in hand, walking through the double doors out onto the practice field for our afternoon practice. One of the quality control coaches taps me on the shoulder and tells me that Bill wants to talk to me in his office.

—Bill?

—Yeah. He's waiting for you.

What could he possibly want to talk about?

I throw my helmet in my locker and walk upstairs. My cleats clack on the linoleum. Heads look up from desks in cubicles to see what beast this way comes. Bill's door is open and he's sitting at his desk. Behind him is a window that looks out the south end of the building. There are framed pictures of his family around the office, and papers stacked neatly on his desk.

—Come in, Nate. Sit down.

I sit. I'm trying to peek at the papers. I wonder if any of them have to do with me. What is it that Bill Walsh reads all day up here?

—Well, I'll get right to it. We've traded you to Denver. As you know, you've been stuck down at the bottom of the depth chart. I asked Coach Erickson if you were going to make the team here. He said no. So I asked his permission to make a few calls on your behalf. I called Denver and Mike Shanahan was interested. He's a great coach. You'll get a fair shot there. I can promise you that. I think this is exactly what you need, Nate. You okay with all of this?

—Yeah, of course.

—I know it's a lot to take in right now but you'll be fine. Your flight leaves in three hours. You better get going. Good luck, Nate.

I thank him for everything, we shake hands, and I'm out the door. I change out of my 49ers gear in the empty locker room and leave. My teammates will come back after practice and my locker will be cleaned out. I will never see them again without a helmet on.

Two hours later I'm at the airport with a duffel bag. I am meat, traded to the highest bidder: the only bidder. Fine, I'll be your meat. I'll be whatever you want me to be. Just give me a helmet.

2

My Life as Randy Moss
(2003)

I arrive at Denver International Airport and am greeted by a driver holding a piece of paper with my name on it. I sit in the backseat and look out the window as he narrates the passing landscape.

—You can't see it now but directly ahead of us are the Rocky Mountains. Beautiful sight when it starts snowing. Usually get our first snow in late September or early October. As you can see, downtown's over thataway.

He points across the passenger seat with his gloved hand.

—But we're headed south of that to Dove Valley. That's where Broncos headquarters are. Boy, Denver is Broncos crazy, I tell ya. I'm not a Broncos fan myself. No offense. Everyone takes football so seriously around here.

—Huh.

—No idea why they built the airport so far away. Kinda makes you feel like you're landing in the middle of the prairie, doesn't it? And I gotta drive up and back and up and back, all day long, forty minutes each way. Anyway, it could be worse, I guess. You see that building?

—Yeah.

—We call that the 'Sore Thumb' building.

—Why's that?

—Because it sticks out like a sore thumb.

We exit the freeway at the Sore Thumb building and go east on Arapahoe Road. It is lined with car dealerships, all bearing the licensed name of Denver's golden child: John Elway Ford, John Elway Toyota, John Elway Honda, etc.

—Wow, you were right. They really do love their Broncos.

—You have no idea.

He drops me off at the Holiday Inn, right up the street from the Broncos facility. I check into my room and look at the clock: it's just after eleven. I sit down on the bed. I am chasing my dream alone.

Nine hours later I sit in my new locker fiddling with my equipment. The Denver Broncos locker room buzzes around me. I am summoned into the training room where I have a brief physical exam with Steve Antonopulos, aka "Greek," the Broncos head trainer, whom I'll come to know as a sometimes not-grumpy bald man with a walrusy mustache. He scribbles his findings and files it with my already growing medical chart: "Physical examination demonstrates a left shoulder that appears stable on exam today after arthroscopic stabilization and some minimal achilles tendinosis. His plan therefore will be for routine foot care, continue shoulder strengthening exercises and treatment as needed. Continue anti-inflammatory medications and treatment in the training room for his left achilles tendon."

I'm given jersey number 14: standard-issue training camp receiver number. The eighties numbers go to active receivers and tight ends. The rest of us get numbers in the teens—the leftovers, basically. I jog out onto the field for morning practice and my new teammates look at me and do a double take. Last night number 14 was a short black guy. Today he is a tall white guy.

Before practice starts I meet my new position coach. Steve Watson, aka "Blade," was a star receiver for the Broncos in the 1980s. He played his entire nine-year career in Denver. Blade is tall and lean with a full head of dark hair and a friendly disposition. Some old-time football players hobble through life and look like they're about to take a knee at any minute. But Blade's still springy and spry. He welcomes me with a handshake and a smile and offers a few words of encouragement. On my way out to the field I meet Coach Shanahan, too. I'm obviously nervous. He's a small man but outsized, with a presence as big as the Rockies. I stammer through a greeting, thank him for the chance, and take a deep breath. I'm going to need all the oxygen I can get.

Halfway through practice I stand on my own with my helmet in my hand, trying to catch my breath after a series of scout team plays. A voice startles me.

—You'll get used to it. It'll take a few weeks but you'll get used to it. No one realizes how bad the altitude change is until they get here. I'm Mike.

Mike Leach: the team's long snapper and a backup tight end, a New Jersey kid and a standout tight end and punter at William & Mary. Coaches often say that the more you can do, the better. Mike took that to heart and learned how to throw a two-handed spiral backward, between his legs, while looking upside down. Thirteen years later and he still has a job in the NFL.

—Hey. I'm Nate.

—Nice to meet you. Just get in last night?

—Yeah.

—Man they got rid of that guy quick. You'll like it here, though. Coach takes good care of us. Let me know if you have any questions about anything. I know how it is arriving in the middle of camp.

After practice the media want to talk to me. Strange. The 49ers media didn't care about the bottom of the roster. The Denver media, I'll learn, care about the whole team.

Adam Schefter, Denver's number one Broncos antagonist, sidles up to me.

—Now, do you know about the history of your number?

—My number? Fourteen?

—Yes.

—No.

—Well, the last two guys to wear it didn't turn out so good around here. Some think it's cursed. Are you concerned about that?

—I am now.

Not only did last night's number 14 disappear like a ghost in the night, but the guy who wore it before him was a dirty word in Denver: Brian Griese. Griese took over for John Elway after Elway won two Super Bowls, dropped the mic at the fifty-yard line, and galloped into the sunset on a white horse. His residue still shines on all things Denver. His name is everywhere: in the newspapers, on the backs of children, on the lips of every talk radio personality in the area, and, of course, on car dealerships. In the three years since he retired, the Broncos have struggled to find his replacement. How do you replace a legend? You don't. But in the NFL, you fucking better! Griese didn't measure up so it was off with his head. Before it stopped rolling, Coach Shanahan signed highly esteemed free agent Jake Plummer, who arrived from Arizona in the off-season. My neophytic opinion after day one is that the locker room has obviously accepted him. Plummer's the guy, part of the crew: no seams or cracks. I want to be a part of it, too.

I change into my new Broncos sweats and realize what was missing in San Francisco. The guys here are enjoying their work. The locker room is lively and loose. Everyone is friends: white guys, black guys, all guys. I didn't realize it at the time, but that was Mike Shanahan's doing. He brought in good personalities, not just good football players. Shannon Sharpe, All-Pro tight end (whom Shanahan had converted from receiver) is a boisterous locker room presence and a beast on the field. His voice carries through the locker room and the laughs follow in his wake.

And the leadership gives every meeting a purpose. In my position group alone are Rod Smith and Ed McCaffrey: two Pro Bowl receivers nearing the end of their careers. The Niners had veteran guys, too. Terrell Owens was an All-Pro and surefire Hall of Famer. But I hardly heard a peep out of him while I was there. Rod's vocal. He is thirty-three years old, six feet tall, and two hundred pounds. He is the consummate professional and he thrives on sharing his knowledge—from coming out of a break on a comeback route, to coming out of a club before the cops come out. He considers it a sin to keep anything inside if he thinks it can improve his team. If he ever notices us doing something wrong, he'll pull us aside and give it to us straight. I decide I'll learn how to be a pro from watching Rod.

But I don't have much time to make an impression: It's already mid-August. Camp's winding down. But I feel fresh and energized. Training camp inevitably becomes mundane and guys get sluggish going through the same routine every day. Changing up the scenery has given me a spark that I know I can use to my advantage.

After a few weeks of good practice, my life in professional football comes down to the last preseason game, at home, against the Seattle Seahawks. The atmosphere at [Insert Corporate Logo Here] Field is impressive, especially compared to dilapidated Candlestick Park. The stadium is full and intimate, the weather crisp and clear. The cheerleaders lithe and sexy. The hot dogs wafting sweet and wonderful all around us.

I play special teams and, on a kickoff, make a tackle, tripping up a returner who was about to bust through the wedge untouched. Then in the fourth quarter I'm wide open in the end zone on offense. Danny Kanell, the backup QB, throws me a high ball. I leap, but it slips off my gloveless fingertips. A defender jumps on top of my body and lands on my shoulder as I reach for the elusive ball in vain. Lucky for me it's my good shoulder. Now it's my bad shoulder, my second bad shoulder. But already I don't care about my body. It was the only ball that came my way and I didn't catch it. I'm the only receiver who doesn't wear gloves. I have never needed them before. But the altitude in Denver makes the ball slicker, drier, and faster through the air. I decide after the game that if I make the team, I will spring for a pair.

That night we go out and get hammered. I'm with Ashley Lelie and Charlie Adams—two fellow receivers—and Kyle Johnson, a fullback. Ashley is a first-round draft pick and a lanky speedster from the University of Hawaii who doesn't take anything very seriously. Charlie is a record-breaking receiver from Hofstra, a friendly, outgoing guy, always upbeat, always smiling. Wherever we go, everyone always knows Charlie. Kyle is from New Jersey and went to Syracuse. He is a thick, powerful, thoughtful man: a philosopher in a warrior's body. He detonates the dumb-jock stereotype with ease. I like my new friends already. It's my first real night out in Denver and it's raining the wavy hard rain of Colorado summer storms. I have a new girlfriend back in California whom I met in the summer. Her name is Alina. We are very excited about each other. We talk on the phone, a lot. We will make it work, regardless of where I am. That was our vow.

But it will prove difficult. The world is ours in Denver. I learn that very quickly. There's never reason to worry: Drink up, young stallion. And keep your wallet in your pocket. Your money is no good in this city. Your dick, however, is another story. Keep that thing ready. You never know when you'll need it.

The next morning I sit on my hotel bed watching the phone. Coach said to be ready for a call between eight and twelve. That's when they'd be doing the cutting. I don't know what to expect. I've practiced hard, made some good catches, played well on special teams. But it's all happened so fast. My phone rings shortly after nine. My gig is up. Time to go back to California and move on, sell insurance, and have sad sex with my girl. But General Manager Ted Sundquist is on the phone saying in fact that I'm not being cut loose. They want me on the practice squad. I clear waivers—the twenty-four-hour period that I'm able to be picked up by another team—then I go upstairs to Ted's office to sign my contract: $4,350 a week. He congratulates me with a handshake. Ted is a former fullback at the Air Force Academy who worked his way up the scouting ladder in Denver. His hair is exquisitely coiffed and he knows it. He was promoted to GM by owner Pat Bowlen two years ago. Ted has an amiable, genuine way about him. I like this Denver place. And now I have a bona fide job in the NFL. Look, Ma, I'm a Denver Bronco.

On Tuesday, our first day off of the regular season, I buy a new Denali. It's a foolish purchase but I can't help myself. I could picture it with my eyes closed as soon as I signed my contract. I love the Denali's angles, the chrome grill. I love the idea that I can buy a giant luxurious machine with only my football skills.

Still, practice squad players have less job security than anyone on the team. They are shuffled around constantly. If someone on the active roster gets hurt, the scrambling and rearranging often squirts a practice squad guy onto the streets. But I had no vehicle, not even my green Civic, and when I looked around at the players' parking lot, I glimpsed the spoils, at minimum, that my talent might afford me. No point in saving every penny. Might as well try to keep up with the Joneses, just this once. And the Denali does make me feel like I have accomplished something. It gives me a tangible reminder of my hard work. Every morning when I jump onto the soft leather seat and turn over that sweet engine, I tell myself that I better have a great day at practice or I won't be making the payments. I'll be "workin' a nine-to-five with a thirty minute," just like our special teams coach Ronnie Bradford says will happen to us if we keep fucking up the plays.

Later that week Rod holds a money meeting for rookies. Rod loves money. He loves making it and seeing it grow. He loves talking about it and he has a wealth of knowledge about everything monetary. All you have to do is ask him a question, sit back, and listen. During the meeting, Rod is stressing the importance of paying attention to what things actually cost. He says that we often have a skewed perspective of the money we are spending because all we are doing is signing our names. Well, he says, if you buy a $50,000 car, it's not just your signature. This, he says, is what you're spending. He reaches into a small black bag and produces five stacks of wrapped cash, ten grand apiece. He drops them on the table with a plop. So that's what fifty grand sounds like. This is the year that the meaning of money will change for me, forever.

There are five of us on the practice squad. Our obligations are simple: practice hard. The NFL workweek starts on Wednesday and ends on game day, with an adjustable Monday schedule and nothing at all on Tuesday. Wednesday, Thursday, and Friday are really all that matter for a practice squad player because those are the days we practice.

My job is to run the opposing team's plays during the week leading up to the game. An assistant coach pores through our opponent's game film and draws up every one of their plays on large cards. He color-codes each skill position and writes the jersey number of each player inside the colored dot that indicates where he lines up. There's no memorizing of plays, no learning of concepts. Just find your colored dot and copy it, shmuck. All week I'm the same dot.

Practice squad players don't travel with the team. It's a strange feeling watching the games on television after the week of practice. I'm a part of the team all week. An important part of the team, too. I prepare our defense to dominate. The better I play, the better they play. I want to make it harder on them in practice than it will be on Sundays. When our DBs play well, I pat myself on the back. When they don't, I take it personally.

Usually when our team is off in another city playing, Charlie and I go to Earl's, our favorite restaurant, to get fed and drunk. Then we saunter downtown and hop around a handful of bars and clubs, dividing our attention between all of the girls who want to hang out with a Bronco. No Broncos around here but us.

—But you're on the practice squad!

—But you're a slut!

And we have an understanding. Alina can feel it with every minute that I don't respond promptly to her texts, don't reassure her that the party's boring, that the girls are ugly, that I'm having a horrible time. I feel her desperation but I'm too weak to be honest with her about what I'm up against: a mob of bloodthirsty jersey chasers.

As the season wears on and injuries stack up, I become more and more anxious to join the active roster. When an active player gets hurt, a practice squad player often takes his place. I want to be on that field "taking live bullets," as we say. Not only do I want live bullets, I want a bigger paycheck. My salary is the lowest in the locker room. I don't think about it early in the season but as the months pass, it's impossible not to. I'm a rookie and so every once in a while Rod has me go upstairs to pick up his check. I can't help but take a peek at it. My eyes nearly pop out of my head. Rod has my $4,350 a week in his couch cushions.

Some time after Thanksgiving, we have our offensive rookie dinner, minus the offensive linemen. I've been hearing stories about the previous year's rookie dinner at Del Frisco's steak house. Clinton Portis had flown in some adult entertainers from Miami to provide flexible, pink comic relief in between bites of filet mignon. The bill was outlandish. The rookies picked up the tab.

Now that C.P. isn't a rookie anymore, he plans to take full advantage of the situation. He brings some friends and they bring some friends. The back room of Del Frisco's is full. Wine and champagne and cognac are flowing like the rivers of Capistrano. Ashley Lelie shows up late and orders two bottles of the priciest wine I've ever seen. The waitress starts to cork one of them and Ashley stops her.

—No! No! Don't open it. I'm taking them home.

Then he leaves with his loot. The bill's over $26,000, split four ways between the rookie offensive skill position players, none of whom was drafted very high. Jake, bless him, feels so bad that he palms me some money to help me pay the bill.

Still, I'm not losing any sleep over money. Football was never about the money to me. It was about competing with the best athletes in the world. And just practicing with them isn't enough. I want to get hit. I mean really hit. I want to hit the ground hard and get up shaking myself off because I think I'm dead. That's the feeling I want. When I was in high school, my friends and I used to have typical teenage stoner conversations. Who would win in a fight between Mike Tyson and two lightweights? What would be the best way to thwart a shark attack? Could I kill a mountain lion with a pocketknife? And would I be able to take a hit from Levon Kirkland?

For my friends and me, Steelers linebacker Levon Kirkland represented the pinnacle of big, scary football players in the mid to late 1990s. They insisted that there was no way I could take a shot from him. I vehemently disagreed. Of course I could! He's only a man, after all. It's football. But there I was stuck with no way to prove them wrong.

Every week of practice, my colored dot means that I'm a different player. My favorite week is when I get to be Randy Moss. Randy's still with the Vikings and is still a badass with some very unorthodox habits. It is ingrained in the mind of football players to go hard, 100 percent on every single play, maximum effort all of the time. So when we watch Randy Moss on film, his tendencies stick out like a dead fish at the aquarium. He has the habit of taking plays off completely. On run plays, he might literally walk off the line of scrimmage, or jog, or skip. He rarely engages the corner in contact on run plays, and generally avoids it altogether. This tendency allows me to do the same throughout the week. I'm a Method actor. I take my role seriously. How seriously? Ask the wives of the players I impersonated.

Randy's inconsistency works to his advantage and allows him to survey the scene, wait for his moment, then attack the defense downfield or across the middle. Football is about angles: linear movement in a contained area coupled with a finite amount of time in which to exploit it. Randy understands that finite time period better than anyone and can narrow the gap between action and reaction because he's really damn fast.

As he lazily skips off the line of scrimmage, everyone else explodes into the play. This forces the defenders to account for both his slowness and everyone else's speed. The defenders pay more attention to the things around them that move faster, and Randy's able to let the play develop in front of him before he joins in, and streaks up the field for an 85-yard touchdown.

Of course, it doesn't always work. You don't even get to try that stuff unless you're Randy Moss, or an actor playing Randy Moss. This week is the most fruitful week of my practice squad days. I catch a million balls: short, intermediate, deep. At the end of the week, Charlie has to pay me twenty dollars. He's a practice squad receiver, too, and we have a running tally every week. Whoever catches the most balls gets paid. Touchdowns are worth two. Big money for a couple of big-time guys.

When the game arrives, and the team ships out to Minneapolis, I'm not in the partying mood. I wake up early on game day and drive to the foothills for a hike and some fresh air. It's a crisp, sunny November day; patches of snow dot the ground and glare white in the sun. I stop next to a creek, find a handful of sticks, and pull out my pocketknife. I whittle off the knots, kneel down, and fashion a water wheel in the shallows. My dad used to do that when my brother and I were kids.

I sit down on the bank and watch the wheel turn with the current. Soon I lie back on a soft patch of dry grass and drift off in the afternoon sun.

The next thing I know, the sun is gone and I'm running through the forest with my knife in hand. I come upon fresh mountain lion tracks that lead to a cave. There's a rustling sound inside and baby paw tracks around the outside of the cave. I pick up a rock and throw it into the cave. Then another. Then I throw a large stick. I taunt the lion, yell at her, insulting her choice of caves.

I see a snake sliding along a fallen branch on a gentle slope toward a dry creek bed. I pick him up by his tail and fling him into the lion's den. Two cubs come bounding out, followed by their mother, who recognizes me as the idiot causing the commotion. She squares me up and bears her teeth.

—Here kitty kitty. Here kitty kitty.

I flash my blade. She squats down and leaps at me with her front legs out and her jaw wide open.

During our stony arguments I could never convince my friends that I'd have the wherewithal to actually, tactically, complete the task. They thought there was no way I could find her neck with a little knife. And besides, her jaws and claws were razor sharp and her hide was too thick to penetrate. She would rip me to shreds.

But at the moment of truth time screeches to a halt. I see each molecule and fiber of both our bodies moving in unison. The will to survive reaches down and inflates me. As she leaps I stand my ground and take a slight step to my right. Using my left forearm as a shield I catch the brunt of her charge and sink the knife into her neck with my right hand. It presses deep into her flesh as if in slow motion. The blade's an extension of my pointed will. She winces and stiffens. Her hot breath hits my face and she thuds to earth.

A cold wind shivers through me. I sit up like a shot and look around. All's calm around me. The only sound is the water wheel gently turning in the shallow creek. I look down at my right hand, quivering and gripped tightly around my folded pocketknife. I hike back to my truck and drive home.

I hear from the radio that we've lost in Minnesota. A predictably erratic and brilliant performance by Randy has done us in. Seconds before halftime he caught a long pass and, as he was being tackled on the twenty-yard line, he pitched it backward to a teammate, who scored as the clock ran out. He ended up with 10 catches for 151 yards. Nothing compared to the week I had.

With four games left in the season we have an injury to one of our active wide receivers, and Charlie and I both cross our fingers. God chooses Charlie. He's activated that week for the game against the Kansas City Chiefs. I stay down on the practice squad. Charlie's pay goes from $4,350 a week to nearly $15,000: minimum wage for an active rookie. I decide to go take a nap in my Denali.

Two weeks later, the team is traveling to play the Colts in Indianapolis. We are 9-5 and have won four of our last five games. It's our biggest game of the season. If we beat them we secure a spot in the playoffs. Blade tells me that if I want to go to the game I can; all I have to do is ask Coach Shanahan. I approach him in the hall before a meeting and ask him if I can go.

—Of course you can, Nate. All you had to do was ask.

On Saturday morning before we leave, Shannon Sharpe addresses the offense. He always gives a talk on Saturday morning before we review the film. Shannon's a three-time Super Bowl winner: two in Denver and one in Baltimore. This is to be his last season in the NFL. He wants one more ring. During his monologue about everyone in the room having a job to do, he says, "Whether you're Shannon Sharpe or you're Nate Jackson, everyone has a role on this team." I'm flattered that I popped into his head, even if it was when he needed the lowest man in the food chain. At least he knows my name.

I get my own room in Indianapolis like everyone else, complete with the two free pay-per-view movies, a staple of Broncos hotel accommodations. I watch porn. No, I don't. Wait, what?

We win the game impressively and clinch a playoff berth. The dome in Indianapolis is usually one of the loudest in the league but it falls completely silent about halfway through the fourth quarter. Peyton Manning rarely lost there. It's a team win that will likely set up a rematch in the first round of the playoffs, after one final game.

The atmosphere is jubilant at work on Wednesday morning. It's Christmas Eve and we're going to the playoffs no matter what. Jake has brought us to the postseason in his first year as our quarterback. Since we would gain nothing from winning our last game in Green Bay, Coach plans to rest some of our key starters. Rod's one of them. They need to activate a receiver to take his place. Blade pulls me aside after morning meetings on Wednesday and tells me the good news.

—Congrats, Nate. You deserve it.

Before practice, I go upstairs to Ted's office and sign a new contract. My practice squad days are over. I'm a member of the fifty-three-man roster. The $4,350 a week is dead forever: chump change for my couch cushions. I'll get my $15,000 for this week and will watch that number rise steadily every year forward. Daddy's got a new pair of shoes for every day of the month.

We have a tradition in my family that goes all the way back to my infancy. I am the youngest of my dad's six children and my mother's two, and every Christmas Eve, my brothers and sisters come to our house with their families and it's one big lovefest. But this year there will be no group hug for me. I'm with my new family. Football takes precedence over everything: even Jesus.

Ed McCaffrey invites me over to his house for Christmas dinner, under one condition: I have to dress up like Santa Claus and play with his kids. Ed is a quirky Stanford grad and is Denver's second-favorite son. When he makes a catch, the crowd chants *E-ddie, E-ddie*. Because of our mutual whiteness and similar size, I was often compared to him coming out of college. I hope to live up to those expectations, as skin-deep as they are.

I show up at the agreed-upon time and meet Eddie at the side of his house, where he has already prepared my costume. He tells me to come to the door in ten minutes. I put on the white beard and the red suit, take a few pulls from Santa's whiskey, and head to the front door.

—Ho! Ho! Ho!

I bellow like a lunatic making minimum wage at the mall. Eddie's wife, Lisa, opens the door with a wink and in I go with my sack of toys that Eddie left me, my deep Santa voice echoing through the house.

—Well hello there young man! What's your name? Ho! Ho! Ho! Have you been a good boy?

And so on. Things are going fine with the youngest two, but the oldest boy, probably eight or nine, stands at a distance regarding me suspiciously. After ten minutes of jolly platitudes I back out the door and head down the path to the side of the house. I change back into my civilian clothes and sit around next to the garage for a while.

Then I reenter through the same door I had exited and am warmly received as if for the first time by all of the house's inhabitants—except for the oldest boy, that clever little buzzkill, who puts the exclamation point on his tiny epiphany: my daddy plays football with Santa Claus.

The game on Sunday is at Lambeau Field in Green Bay. Though the game doesn't matter for us in any real way, they need a win to make the playoffs. They also need some help from the Cardinals. The Vikings are ahead of the Packers in the standings, but if the Vikings lose and the Packers win, they're in. We drive through the neighborhoods of Green Bay on the way to the stadium and I'm struck by the surrealism of the moment. I'm on a bus with my NFL team in Wisconsin on our way to play the Packers. The entire town of Green Bay *is* the Packers. On every lawn there are signs and banners and parties and barbecues and happy people bursting at the seams. With wide, unthreatening grins, they drink responsibly and politely urge us to go fuck ourselves. Lambeau Field lies at the edge of what appears to be a typical midwestern suburban community, unlike most other stadiums, which are built in downtown or in industrial areas. In Green Bay, the stadium has the feel of being a park at the end of the street. We are riding along and I'm looking into living room windows, then all of a sudden, like a Mecca of cheese:

Lambeau.

I find my locker and sit down. There is a program for the game on my chair and I thumb through it while listening to my mood music. I look for the cheerleader photos in the back of the program, a road-game ritual of mine, and am disappointed to find that the Packers don't have any. But I see my name on the roster list, and here I am, sitting at my locker: a man alive inside a dream.

I jog out of the tunnel and take it all in. A security guard in a yellow jacket smiles and wishes me luck. It's a crisp, dry night. Once the game starts I stand at attention next to Blade and when someone needs a rest I run onto the field and into the huddle. Jake's resting, too, and since Steve Beuerlein snapped his finger in half a few weeks earlier, Danny Kanell and Jarious Jackson are sharing the duties at quarterback. The Packers kick our ass. They're playing for something and we aren't. But when I'm on the field I feel calm. I see things happen in slow motion. I'm comfortable. It is still just football. People always question whether a guy can perform "when the lights come on," when the moment is big. But that's bogus. The magnitude of a game is manufactured by those who sell it, not by those who play it. The lights are always on.

The next week we get back to work preparing for our playoff run. But in the opening round, we travel back to Indy and get rolled. Peyton's flawless. Eddie Mac had a head injury from the Packer game so I'm suited up again. I stand on the sidelines until the last drive. We are running out the clock with some standard inside runs. Most everyone out here is playing patty-cake and waiting for the clock to hit zeroes: the Colts have another game to prepare for and we have the off-season waiting for us. Blade puts me in for the last four plays and I run around like a crazed jackal. All of them are knockdown blocks or close to it. I want blood. I want to taste the iron on my tongue as I rip the flesh from a safety's bones and play Hacky Sack with his testicles. Everyone looks at me like I'm an idiot. The free safety yells at me after I crush him with a borderline illegal block. But I don't care. It's my playoffs, too. The clock empties and our season ends. And the only blood I've tasted is my own, in the form of two vicious carpet burns from the NFL's last proprietor of AstroTurf hell. For the next month I wake up sticking to my sheets.

The week of the Super Bowl, Charlie and I fly to Houston to pick up our Super Bowl tickets. NFL players have the option of purchasing two tickets at face value, but for some reason they make the rookies pick them up in the Super Bowl city. Veterans can pick up the tickets in their home cities. The markup for Super Bowl tickets is obscene, so we take a business trip to Texas to purchase our tickets at face value before selling them at a "significant markup." The ticket scalping underworld is a breeze once you're in. True market value reveals itself in back parking lots and dark alleys.

We go to the designated hotel and get our tickets, then we meet our handler in a different hotel parking lot. He gives us a wad of cash and we each hand over two pieces of cardboard. Paper for paper, the American dream unfolds. We book a room in a cheap motel and float from party to party, denied entry at nearly every one, and settle for a gentlemen's club, where I fall into a deep conversation with a New Orleans dancer who has come to town to cash in on the Super Bowl muscle. I flex my practice squad muscles for her. She is not impressed.

3

Nein Lives
(2004)

My phone rings. The caller ID tells me it's coming from the Broncos facility.

—Yellow.

—Hey, Nate, it's Blade. How's it going?

—Hey, Blade. All is well. Driving through the Rocky Mountains right now.

—Ah, headed home, eh? That's great. Enjoying your off-season?

—Yeah, so far. What's up with you?

—Ahh, you know how it is. It's off-season for you guys but not for us. We're in here burning the midnight oil. Anyway, the reason I'm calling is that we've discussed it as a staff and we think you'd really benefit from heading over to NFL Europe next month. I know what you're probably thinking, Nate, but it would be great. You'd get some game action under your belt and you'd have a great time out there, Nate, you really would. And you'd be back in time for our last few minicamps. We really think this will be great for you. So what do you think?

I don't think.

—Yeah, Blade. Let's do it.

Great. I had a feeling I might be getting that call. NFL Europe is a supplemental league owned by the NFL and used as a de facto farm system. There are six teams: the Scottish Claymores, Berlin Thunder, Amsterdam Admirals, Cologne Centurions, Frankfurt Galaxy, and Rhein Fire. The NFL's off-season is NFL Europe's in-season, so practice squad players like me are often sent to NFL Europe to develop. Charlie played in NFL Europe for the Rhein Fire last year and he's told me stories. Football in Germany? Man that must suck. But he loved it. And he was the one who warned me that I might be going. Nah, I thought. Not me. I'm going home in my Denali.

And I do go home in my Denali, but only for a few weeks. Yet it's plenty of time to see that things have changed for me back in San Jose. All of my friends will drink for free tonight, here at our neighborhood bar, the same bar we've been coming to since we were teenagers. Now my money is no good here. Now the girls are lining up. Now people are offering me rides home. Now I'm a Denver Bronco.

Along with the newfound adulation comes a new responsibility: my urine no longer belongs to me. While I'm home I get a call on my cell phone from the Pee Man. I'm on the list. But I'm not in Denver. I'm at home with my family. That's okay, he says, we have someone out there we can use. Northern California's regional Pee Man meets me in a parking lot and follows me to my parents' house. I introduce him to Mom and Dad on our way to the bathroom. I pee in a cup and hand it to him. He squats in the hallway and pours it into two sample cups. He caps each of them and seals them both in a box. I initial and sign everything and he leaves, nodding to my confused parents as he walks out the front door holding a box of my piss.

I'm allocated to the Rhein Fire in Düsseldorf, Germany. But our camp is in Tampa. We arrive in late February at our expansive hotel compound just outside of town.

Wide receiver Adam Herzing is my roommate. We know of each other from back home in San Jose. He's a year younger than me. We went to the same middle school and rival high schools. We both went to Cal Poly and have the same agent—good ol' Ryan Tollner. We're both six foot three, white, and love Cheerios. But we had never met.

We become fast buddies, along with Greg Zolman, a six foot two lefty quarterback from Vanderbilt, whose room is across the hall. Greg and Adam know each other from time spent with the Colts. Greg shattered Vandy's passing records and is now competing with NFL Europe golden child Chad Hutchinson, allocated by the Cowboys and serving as the poster boy for NFL Europe's "Here's a name you might recognize!" campaign. But everyone has a history. Everyone has expectations. Everyone's name is recognizable somewhere.

Before we can practice, we have to go through physicals and introduction meetings. The first meeting is in a banquet hall and led by the commissioner of the league. Then, after a few more forgettable presentations, we are greeted by our German sensei, Markus. In his polished English, touched slightly with a Teutonic accent, he sets the terms.

—Ze food vill be different, ze langvich, ze transportation, ze customs, ze people. Everyzing vill be different. You muss know zat.

Yawn goes the crowd. But I'm intrigued. Sure, NFL Europe isn't how I expected to spend my first off-season. And sure, the money is shit compared to the NFL. We'll make $600 a week. But I'm headed to Europe to play the game I love.

After Markus finishes his spiel, we have our physicals: six teams' worth of football players to inspect and all of them with a lifetime of injuries to identify and document. NFL Europe contracted HealthSouth, a medical group based in Birmingham, Alabama, to oversee all of the major bodily issues. Bumps and bruises will be treated by team trainers in Europe but anything more serious will get you a one-way ticket to Birmingham. HealthSouth's head trainer is Mayfield Armstrong. I hear the voice before I see the man. As the long line of players approaches the door to the banquet room, Mayfield holds court inside. Gregarious but firmly pointed, he shouts instructions.

—C'mon, Jimmy! You wanna play professional football I'ma hafta see ya at least *try* to touch ya toes! Is that s'far's you can go, big boy?

—Now I ain't the smartest man alive, Julie, but it says here this man just came off an ACL surgery. Looks pretty good to me!

I get to the front of the line and sit down in front of Mayfield. He's a husky middle-aged man, clean-shaven with graying hair and a fierce twinkle in his eye. He looks at my file.

—All right, Mr. Nate, says here you had a shoulder operation last year.

—A year and a half ago.

—How's it feeling?

—Great, it's great.

—Well show me then. Can you do some push-ups?

—Really?

—Yes, sir. Really.

I drop and do ten push-ups.

—Good. Can you give me a little clap at the top of it?

I do two with a clap. He scribbles some notes.

—Well, all right, Nate. You look healthy to me. Good luck over there.

We shake hands.

—Don't let me see you again until the exit physical.

Head coach Pete Kuharchek has our ear.

—You guys have a great opportunity. Some of you are already on NFL teams. Some of you are trying to get on NFL teams. But we're all here now. And there are some damn good football players in this room, guys. We have a lot of talent, the best talent in this league. But it's up to you to make the most of it. Things are going to be different over there, guys. Okay? You are going to have to adjust. If you're expecting everything to be like the NFL, you're going to be disappointed. Keep an open mind, guys, and roll with the punches. The football part will be exactly what you're used to. Everything else, well, we simply don't have the budget, okay? Our motto is: Be flexible. Okay? Be . . . flexible.

The 2004 Rhein Fire sit at attention, watching our new coach dig into his opening monologue. Bent forward at the neck, his spine is a taut bow ready to fire an arrow at whoever might pursue him, a shot that would snap him into a posture two inches taller. He stalks his ten feet of real estate at the front of the room and shakes his head almost imperceptibly with each point of emphasis, his lower lip weighed down by a pond of saliva formed by the angle of the bow. Pete is a veteran coach who's never been on an NFL sideline. He has, however, taken the Rhein Fire to back-to-back championship games. They lost both times.

—Practice is going to be physical, guys. We're going to hit each other. We're going to be violent. We are going to bully people out there. And it starts with training camp. There's no way around it. We are going to work our asses off while we're here. I promise you, none of these teams are going to outwork us. We'll be in Germany before we know it. But right now we have work to get done. So bring your hard hat and your lunch pail to work every day, men. This isn't a vacation.

Great. Pete wants us to kill each other.

Practice is at a high school in Clearwater Beach. From our hotel it's a forty-minute drive across the 60, a toothpick bridge suspended over the crystal waters of the Gulf of Mexico. To make sure we are all frothing at the mouth to hit someone, Coach splits up the offense and defense for the first few days of practice. We practice the plays on our own, by ourselves, with no defense. They do the same on their end. It is boring and gets us all riled up. Football players are conditioned for violence. We are at home in the melee. We may have moments of quiet reservation and doubt when lying on our living room couches, but on the field we are pulled toward the mayhem. The feel of the helmet and shoulder pads, the sound of the whistle, the taste of the mouthpiece, the smell of grass and sweat: sacraments for bloodshed.

But the only interaction we have with the defense is in the locker room and on the bus, and since we aren't getting to know each other on the field, the locker room and bus are quiet. We are strangers. On the day we are to finally practice as a team, the tension is high. Our sacraments have been dangled in front of our noses but we've been kept in cages. Just before they unlock the doors everybody is talking shit to each other from across the field. It feels like we are going to brawl.

The first thing we do is a passing drill called seven-on-seven that is designed to work on pass plays only, without any linemen getting in the way. The defense plays coverage and tries to prevent the passes from being completed. But they don't hit the receivers. They protect their vulnerable teammates.

On one of the first plays of seven-on-seven, I catch a pass across the middle, turn up field, three, four steps, and am cracked hard from the side by a safety. I pop up and look at Whiskey Pete. This is the moment that sets the precedent, the moment where coach says . . .

—What the *fuck* is that?! We don't do that shit around here! You got it? Does everybody get it? Save that for game day. We're on the same fucking team, guys. Protect each other!

But he just stands there watching us through his eyebrows, lip pond glistening. That's what he wants. Well all right then. That's what he'll get.

The next hour and a half is a bloodbath. Bodies are flying and helmets are cracking in the Florida sunshine. Must . . . impress . . . the . . . coaches. Smack! The dreams of the father! Smack! The *American* dream! Crack! C'mon, *boy*! Whammo!

Thirty minutes later, on a routine run play, I size up the strong safety for a block. He comes at me in kill mode. We meet solid: helmet to helmet and chest to chest. But also knee to knee. The bursa sac on my left knee bursts. Fluid rushes to cover my patella. He isn't so lucky. He yelps and falls at my feet. Our best defensive player is done for the season with a torn ACL. Are you not entertained?

After the first practice, things settle down. Now we know each other. Next week, we practice against one of the other teams. Fresh meat. We are doing one-on-ones against their cornerbacks, and their receivers are doing one-on-ones against our cornerbacks. They run a route; we run a route. Pride is on the line. The shit-talking is constant.

Late in the drill, I line up to run a slant. A corner steps out to cover me. He squats inches from my face and mumbles something about handcuffs. He's short, even for a defensive back, and his lowness to the ground forces me to lower my stance to improve my leverage. I shoot off the line and engage him with my hands, then push off and break to the inside, just in time to see the ball soar over my head. I feel a twinge in my pinkie and look down at it. It's sticking out sideways and down toward my wrist at an acute angle. I hold it up in front of my face. Not much of a painful feeling. No feeling, really. I take off my glove. It looks much more real without the glove. I walk my pinkie over to the trainer.

—*Mmmm.* That's dislocated, Nate. Here.

He grabs my pinkie and yanks. It slides back into place without a whisper. I reglove my hand, tape the pinkie to the ring finger, and I'm back to practice.

After practice, though, the finger isn't acting like a reduced dislocation. It hurts. A lot. We X-ray it. It's broken. The X-ray looks like someone has taken a ball-peen hammer to my finger. Shards of slivered bone surround a prominent shark-tooth fragment just below the second knuckle.

They decide to take a closer look at it in Birmingham. The next morning I fly to Alabama. The bursa sac in my knee, manageable at sea level, fills up with fluid on the airplane. I hobble in to see Mayfield.

—Nate! What the hell are you limping for? I thought it was your finger!

—It is, Mayfield. I'm just sore, that's all.

—Shit, Nate. Who ain't?!

They keep me there for four days. They are trying to figure out whether to operate. But there's not much rehab to be done on a shattered pinkie. I spend most of my time wandering around the hospital and flirting with the HealthSouth receptionist. She's a cute, brunette southern girl in business attire with eyes screaming "get me out of here." At every door opening, every phone ringing, every new set of footsteps, she perks up and shoots her flare. She isn't going to miss her chance.

On my last night in Birmingham we go to a movie together and talk. I talk about my girlfriend. She talks about her boyfriend. Both of us are unsure of whatever this is. Unsure of everything. She is starting to realize that she may never make it out of Birmingham. I'm starting to wonder if either of us should bother trying.

Before I leave the next day, Mayfield makes me a pinkie splint.

—All right, Nate, now this should help. But it's still gonna hurt. Shit, you know that. That ain't nothin' new. But I'm serious this time, Nate. Don't let me see you back here. You got it?

—Yeah I got it, Mayfield. Thanks for the help.

I'm on the field for practice that afternoon with my pinkie splint and my knee brace, gimping from another flight-induced swelling. But I'm happy to be back with my teammates. Our receiver core is getting tight. Aside from Adam, there's Shockmain Davis. Willie Quinnie. Chris Leiss. Bosley Allen. Jon Olinger.

The day after I get back from Birmingham, we have a scrimmage against the Amsterdam Admirals. I'm very tired in warm-ups. I feel out of shape from my four days in Alabama. My receiver coach doesn't put me in until the end of the scrimmage. A few plays after I enter the game I catch a 60-yard touchdown from Greg on a blown Cover 2, nearly hyperventilating in the end zone. Well all right. Football is easy. Just throw me the ball.

A wide receiver can only catch what is thrown to him. And it's never up to him. He must run his route and hope. My time spent in the NFL will be full of this hope. I will run every route with gusto, expecting to turn and see the ball spiraling toward me. But it will rarely happen. And with every route I run, beating the world-class athlete being paid to cover me, and being rewarded only by the defeated look in his eyes, a small piece of my football idealism will die. I want the ball. Always. An effortless harmony of quarterback and receiver is a beautiful thing. All is right in the world with Greg Zolman at the helm.

A few days later, after a month of training camp, we have our last team meeting before packing up and heading to Germany. Coach goes over everything again: Ze food, ze buses, and adapting to ze unknown. Then one of the few returning players from the previous year's squad speaks up.

—Yeah, fellas, real quick. Just want to let y'all know, they ain't got no Magnums over there so bring your own rubbers. And bring a lot. You don't want to get caught up.

Advice well received by the team. This will be our Magnum Opus. Someone brings a duffel bag full. *Ich bin ein Düsseldorfer.*

After several long, cramped flights, layovers, and buses, we pull in to our new home in Düsseldorf. The Relexa Hotel. It's a seven-story building on the outskirts of town. The hotel is nice and clean and we all have our own rooms. It's the end of March. We have a week and a half to practice and get used to our surroundings before our first game.

A few days later there is a pep rally in the city's main square. We pull up in our buses and parade onto a stage where Markus works the crowd of a few hundred into a polite frenzy. Frothy cups of good beer tilt at the slight angle of almost drunk and apparently happy. Their enthusiasm surprises me. I hadn't expected the Germans to support the NFL's attempt to make people love the *other* football. But from a dirty seed sprouts beauty. Someone hands me the microphone while we stand onstage. I do my best hype-man impersonation.

—Alo everybody! Are you all having a good time?!

—Ja! Ja!

—What's that? I can't hear you!

—*Ja! Ja!*

—All right! When I say 'Rhein,' you say 'Fire'! Rhein!

—Fye-a!

—Rhein!

—Fye-a!

Then someone snatches the mike and it's on to the next hype man. Then we are ushered offstage and back onto the buses, creeping through a throng of boisterous Germans who have gathered to wave us on.

Our first game is at home against the Cologne Centurions. The stadium is state-of-the-art, featuring a retractable roof and a field that can be rolled entirely outside so as to receive more sunlight, or something. The Arizona Cardinals stadium has the same technology. There are twenty-five thousand fans or so and all of them wear whistles around their necks. They blow them all game, rendering the referee's whistle mute and the concept of "play the whistle" dumb. Be flexible.

I'm on the front line of the kickoff return team. The opening kickoff of our season soars through the air. I turn and run to my landmark, pivot, size up my block, and engage him. Rober Freeman, our kick returner, weaves through the wedge and runs past me with the ball in his hands on his way to the end zone. Touchdown! Touchdown! Twenty-five thousand whistles.

After a touchdown by Cologne, we line up for our second kickoff return. This time Shockmain receives it and shoots past us all the way to the house. I leave my man and follow Shock to the end zone as twenty-five thousand Germans lose their shit, again. This is awesome! This is Germany.

After that, the game settles down. At the start of the second half I am split wide to the right. Greg is in at quarterback. He gives me a look at the line of scrimmage. I run a fade route and he lofts it up: slightly underthrown, just how I like it. I slow up and leap at the last moment. The cornerback has me covered but my jump takes away his advantage. I lose the ball in the lights. I stick my hands out where I think it will come down. It lands in my basket and I squeeze it into my body. As my feet hit the ground the free safety pops me under the chin. But he doesn't bring his lunch pail with him. I bounce off his hit and gather myself, then head up the sideline. The cornerback dives at me. I pirouette and shake him off, heading up the sideline again. The safety who missed me the first time catches me from behind and latches on to my waist. I drag him another ten yards before his buddy jumps on my back and drops me at the five-yard line. *Ja! Ja! Ja!* We score on the next play.

It's a tight game. On our first drive of the fourth quarter, I'm split wide right from deep near our own end zone. I run a five-yard hitch and wait for the ball. Chad Hutchinson is in at QB. He throws the ball over the middle but loses control of it and it dribbles off in front of him. No one knows if it is a fumble or an incomplete pass. And the whistle won't tell us. I run toward the rolling ball and pull up when I realize the play is dead. But not everyone gets the memo. A defensive lineman flies in, overshoots the ball, and lands on my good knee. Pop. No more good knee.

I fall to the ground and grab my leg. It's such a loud pop in my head that I expect a bone to be sticking out. I pull down my sock.

Nothing. Clean leg.

I stand up and walk to the sideline. I tell my trainer something popped in my knee. I try jogging around to shake off whatever just bit me. Unshakable is the phantom of truth. It's no good. I sit on the bench and seethe. We win the game by a point. Afterward I go to the hospital for an MRI. According to the German doctor who reads the MRI, my medial collateral ligament is torn.

—Zat pop you heard vas your ligament tearing, right here.

He points to the apparently abnormal image on his screen, string-cheese-splayed fibers.

—Ze recovery depends on if your doctors decide you need surgery.

—My doctors? You're not my doctor?

—Nein.

—I have nine doctors?

This will apparently fall to HealthSouth. I'll wait until they confer. I go back to the Relexa Hotel and flop onto the bed. It seems that every time I get hurt it's on a play that feels wrong from the start. The finger, the bursa sac in my knee, and now this one. All of them flukes. And my two shoulder dislocations in college were the same: stupid plays that never should have happened. Either I'm rehabbing here in Germany, watching my teammates play, or I'm getting back on a plane to Alabama. Neither appeals.

I pick up the phone to call Alina back home. I need the reassuring voice of my woman to tell me everything is okay.

—Hello?

—Hey.

—Hiii.

—What are you doing?—Me.

—Uhh, nothing. I'm in a cab.—Her.

—Going where?—Me.

—Going home.

—Huh? Home from where?

It is Sunday, 10:30 a.m. in California.

—Ugh, you don't want to know.

—Ugh, yes I do.

— . . . Vince Vaughn's house.

—Vince Vaughn? Why?

—I don't *know*. I'm *so* annoyed right now.

—You slept there?

—Yes. On the couch. Amy hooked up with him.

—Why were you there?

—We were hanging out with him at a club and he said he was having an after-party at his house so we got in his car with him and went back to his house and no one ever showed up. It was just me and Amy and Vince.

—Great party.

—I'm sorry, babe. I didn't know.

—You didn't know what?

—I don't know. It was just stupid.

—And you slept on the couch?

—Yes, baby. I promise.

—Whatever . . . I tore my MCL.

—Aw, *baby*!

(Fuck Vince Vaughn.)

The next day I'm back in Birmingham and back with Mayfield.

—Well, Nate, you want the good news or the bad news?

—Isn't it all bad news?

—Oh c'mon, Nate. Don't get down on me now.

—Bad news.

—Okay. Bad news is, you're gonna be here for a while. We gotta get that thing right and it ain't gonna happen overnight. Good news is, you don't need surgery. That thing'll heal on its own, but we gotta stay on top of it. And the harder you work, the faster you'll get your ass out of here. You look around while you're here, Nate. You're gonna notice some things real quick. One is, there's some sorry sons of bitches around here feelin' *extra* sorry for them*selves*, moping around, going through the motions and ain't gettin' shit done. Shit, Nate, there's players that's been here for over a year. Imagine that, Nate. A *year*! They get so down on themselves that they can't heal. And you know you can't leave until I clear you medically. Make sure you ain't that guy, okay? I know you ain't that guy, Nate, but make *sure* you ain't, you got me?

—Yeah, I got you. I love it here and all, but I gotta get back to Deutschland.

Inordinate pause.

—Germany.

—Shit, I know that, Nate.

• • •

I love Mayfield's enthusiasm. And I promise myself I will adopt his approach: Stay positive. Stay motivated. Every day has a purpose. But it's easy to start feeling sorry for myself in Birmingham. We stay at a Shoney's Inn in a nondescript commercial neighborhood south of the city, across from a U.S. Treasury office, a Dollar General, a handful of other depressing hotels, and an animal hospital. Unlike most Shoney's this one doesn't have a restaurant attached. Instead we have a shuttle service to take us to our meals. Breakfast and lunch we eat in the hospital cafeteria.

Years later, when I'll close my eyes and picture this city, I'll see an overweight woman walking slowly across a street as I sit in the passenger seat of the Shoney's Inn shuttle and wait. At the wheel is Catman. In the back of the shuttle are seven more hungry, injured football players. It's dinnertime and Catman is our ride. The most energetic man in all of Alabama, Catman is one of three Shoney's Inn shuttle drivers. He's a military veteran, maybe, in his forties or fifties, or sixties, with fading tattoos on his forearms and long gray hair slicked back. He weighs 120 pounds and has three prominent teeth, all on the bottom row, all abnormally long and knifing up toward his nose. He got his name because he meows like a cat. He brings his hands to his mouth and twirls his hips while staring down the object of his feline affection. Catman is my support system in Alabama. When things get weird, he'll be there to let me know: That ain't weird, *this* is weird.

—I haven't been with a woman for so long the crack of dawn makes me horny. Meoooow!

—More, Catman! More!

He's an old soul in a new world, a weak world, a humorless world with no sense of adventure. He's seen things and been places. Or he's seen nothing and been nowhere. It doesn't matter. His spirit is on fire. Everyone knows him everywhere we ride in our shuttle.

—Hey, *Catman*.

Tension is high in Birmingham and Catman is only trying to lighten the mood. Mayfield is right. I look around and I see some sorry sons of bitches, feeling *extra* sorry for themselves. I meet a guy who's been here for over a year. He's still on crutches. I meet guys who are moving in on a year. All of them have one thing in common: "Fuck *this* shit." They are folding in on themselves. Shrinking to meet their beaten wills. Injuries in football are common, but being literally shipped off to exile after being injured isn't.

As the weeks pass, my knee gets stronger. But it's still weak. I feign perfect health and pepper Mayfield with questions about my release. When, Mayfield? When? I feel great!

To fill the time, I go to an office across from HealthSouth where we're allowed to use the computers and the Internet. I'm writing a weekly journal about my Europe experience for the Denver Broncos website (which is how I found out, in Tampa, that Ed McCaffrey and Shannon Sharpe have both decided to hang up their cleats for good). The home page today is promoting the new draft class: two new receivers, Darius Watts in round two and Triandos Luke in round six. I try not to think about it. But two drafted receivers means two less spots available for me.

Our veteran backup Steve Beuerlein has also retired in the offseason, so Coach has drafted a pair of rookie quarterbacks in the seventh round: Matt Mauck and Bradlee Van Pelt. Matt's a cerebral, down to earth, mechanically sound, prototypical quarterback from LSU. He won a national championship a few months earlier. Bradlee's a free-spirited renegade quarterback: a running back with a cannon and a thirst for life. He played at Colorado State and has a cult following in the area, among them Pat Bowlen, who urged Coach Shanahan to take a chance on the California hippie who rode his skateboard to class barefoot and excited the crowds with his erratically brilliant performances.

But I can't think about those guys right now. I have to get back on the field, show my coaches that I'm better than them. Sitting in the HealthSouth cafeteria isn't helping anything. I have week five against the Scottish Claymores in my sights. The game is in Glasgow and I have to be there.

I am Scottish. My grandfather was born and raised in Glasgow. He was a musician and had seven children: six boys and a girl. The girl, Mary, is the oldest, followed by my father and five more boys. My grandfather died of a stomach ulcer when my father was fifteen, in 1944. My grandmother raised the seven children by herself in Wenatchee, Washington. After they were all up and out of the house, she moved to Scotland alone and lived there for two years in a flat. Then she moved back to the States and settled in San Francisco. San Jose is an hour from San Francisco and we went to see her often when I was a boy. We called her San Fran Gran. She died at ninety-one, when I was in middle school.

Years later my older brother Tom and my father took a trip to Scotland together. Then Tom studied in Glasgow for a semester when he was in college. But I have never been. So I circled that game on the calendar, and when I showed the schedule to my family we all decided that they would make the trip across the Atlantic and meet me in Scotland. They planned the trip. Bought the tickets. Booked the hotels. Then I got hurt.

And no amount of false enthusiasm will make it better. As sympathetic as Mayfield is, he doesn't care about Scotland: he cares about my knee. I call my father and tell him that I'm not going to be there, that I'm sorry. He tells me don't be sorry, son. We love you. We're proud of you. They go anyway and watch the Fire play the Claymores in our homeland. I watch the game alone at a sports bar in Alabama and get drunk. After it's over I call the Shoney's and Catman picks me up.

—What did the rooster say to the screwdriver?

—Not tonight, Catman.

—Oh, all right, Nate. All right.

The next day I become violently ill and shit and puke for twenty-four hours. I lose ten pounds and can't go into rehab. I lie in my musty room festering like a wound in the heavy air of vomit and excrement and roll myself up into a sheet burrito and pray for the end of everything. I want the screen to go blank and end my misery once and for all. I sit on the toilet with a bucket in my hand, staring at my face in the full-length mirror hitched stupidly to the bathroom door. I don't recognize myself. I'm a ghost, limping, bleeding, crawling after the sunset. Everything is wrong here. I'm dying. But turn around, boy. The sun also rises. Today makes unthinkable the thoughts of yesterday. The bug that squirmed in my body and tried to kill me was killed instead. I lick the blood from the blade and kick the corpse into a shallow grave as I step from my room at last.

—God damn it, boy. You look like shit.

Mayfield.

—You should have seen me yesterday.

—Couldn't have been much worse than today! Go and get some food in you. We'll get to your rehab later. You need to get your weight up before I can clear you.

I spend the next three days in the HealthSouth cafeteria and the following day I'm on my way back to Germany. Mayfield has cleared me after a solid performance on the Biodex machine and a solid enough performance in the running portion of the evaluation. He tells me I better run like a scalded dog if I am going to get clearance. I do my best scalded-dog impersonation and all is right in the world. My knee isn't really healed, but who cares? Nothing ever really heals. Not in football. Not in anything. I can deal with whatever not being 100 percent means but I can't take another week in Birmingham. Things have gotten too heavy.

I'm convinced that the Shoney's is haunted. My dreams are so intense I wake up exhausted. My movie date grows hostile toward me because of I don't know what. Catman asks me if I can buy him a Snickers. Then he asks if I can give him money for his bills. A few

days later he gets fired. Then rehired. Fights are breaking out between injured players in the back of the shuttle on the way to dinner,
in the lobby of the hotel, at HealthSouth. Everything is in shambles.
And it is not going to get any better. The Shoney's Inn is firmly plant-
ed in the Birmingham dirt and NFL Europe injured life descends
upon it like a plague. Mayfield has given me my freedom just in time.

I arrive back in Germany to a different hotel and a far less enthusi-
astic team. We are 2-3 and a week earlier, had left the Relexa Hotel
because of a convention in town. We will stay at a hotel outside Düsseldorf in Hamm for another week, then back to the Relexa for the
last three weeks of the season. At the airport in Alabama I pick up
some magazines to keep me company through the no-doubt sleepless few nights I will endure upon my jet-lagged arrival in Germany.

One is a *Playboy*. The cover girl is a Japanese woman named
Hiromi. The liquid curve of her body, the obvious softness of her
skin, her sweet smile and her raven hair: she jumps into the third
dimension and sits down next to me and we watch late-night German television together for three nights. Late-night TV in Germany
could be a naked woman holding a basketball and dancing next to a
Cessna airplane, or a slapstick talk show, or three grown men wearing only socks and tennis shoes playing a two-on-one tennis match
over house music. You never know.

Hiromi and I laugh for hours until we finally doze off into the
half-dream, half-hallucinatory vacuum trance that seizes the jet-
lagged world traveler upon arrival. The dripping of my bathroom
faucet careens off the walls of a vibrant mental cave and twists the
dial on an ever-expanding ghost hunt, soliciting the expertise of
myriad dust mites and molecular sponge-bath vermin to make my
case before a stubborn queen. I've traveled all this way, my lady, to
bring you this: I reach into my rucksack and pull from it a single violet stone: an amethyst. It twists in my fingers and catches the light
just as—Bzzzzzzzzz. Alarm clock. The buzzing is in German and

the queen has vanished. Hiromi is gone, too. It's time to go to work.

I arrive in the middle of the week. Most of my teammates have already fallen out of love with Germany, and the fact that we have a losing record makes it easy to start counting down the days until it's all over. I try to keep things in perspective for them by explaining the situation in Birmingham. I do the catcall and everything.

During a break from meetings the day before our game, Greg tells me there is some tension on the team after a recent incident at the hotel. There is a group of defensive players who played dominoes every night in the common area of our floor at the Relexa. Greg's room opened up to that common area and every night, according to Greg, they slapped the dominoes down with gusto and laughed loud and made such a commotion that Greg was having a hard time sleeping. In fact, said Greg, he wasn't sleeping at all. The rhythmic slapping of dominoes and laughter brought him to a boil. And repeated respectful pajama'd pleas to please quiet down and please just *put* the domino down instead of *slamming* it down because try try try as I have I cannot sleep when you guys are playing this loudly. It all went unheeded.

Then Greg's girlfriend, Alissa, came to town. It's one thing to deal with hell privately, but when your woman comes around and makes you feel like a punk for letting it happen, something has to be done. As they lay quietly in the dark, taunted by the clacking ivory, Alissa incited a riot.

—Say something, Greg.

—Believe me I have. They don't care.

—Can't you tell your coach?

—I'm not going to run and tell Coach.

—*Well.* You have to do *something.*

—What do you want me to do? I have to live with them.

—If you won't do something, I will.

After dinner the next night, like every other night, the domino crew took their places, washed the bones, and plucked their seven apiece. Only it was three sevens and a six. One bone was missing.

The whole skeleton was useless. Thus ensued the world's most frantic Easter egg hunt. It's the small things in Düsseldorf that allow men to Be Flexible without Going Postal. Abdual, the ringleader of the domino squad, soon came around to the idea that the game was sabotaged. And it didn't take long to come up with a prime suspect: the white guy in room 207.

—Yo, man, you know what happened to our domino?

—Huh? Domino? What do you mean?

—You know what I mean, Greg. Someone took one of our dominoes.

—No. No idea. Are you sure you didn't lose it?

—Naw, man, someone took it. You *sure* it wasn't you?

—I wouldn't do that, man. C'mon!

Propelled by the scent of Caucasian deceit, Abdual convinced the Relexa manager to let him review the security tapes.

Domino, mothafucka!

Sweet little Alissa, wearing a hooded jumpsuit, tiptoed into the hallway, glanced left and right as if crossing a dangerous street, stepped to the gaming table, and plucked the double five. She curled it in her fist and slid back into Greg's room. But the eye in the sky does not lie. Armed with the visual evidence, Abdual returned to Greg's room.

—All right, Greg, we checked the tape and we know you have it. We want it back.

—Yes, I have it and I'm not giving it to you.

—Dawg, you stole it. Coach is going to send you home for this shit.

—You've been stealing my sleep, *dawg.* Coach is sending *you* home!

Back and forth they went until eventually cooler heads prevailed. Greg returned the domino and Abdual promised to try to tone it down. But they are still in the trial run portion of the tone-it-down phase, and things are still testy.

But that's not the only reason things are testy. Pete still has them beating the shit out of each other. Guys are tired and sore and all of that work isn't translating to wins on game day. That's when the rah-rah stuff can start to work against a coach. During training camp, we were the only team that practiced twice a day in full pads. The other teams were done by 1 p.m., hanging out by the pool and laughing at us as we limped off to another violent practice.

—No one is going to outwork us!

Well, fuck, Pete, what's the point of working so damn hard if the dudes who were hanging out at the pool are beating us?

But he's trying to turn us into a good team. He wants us to win. And his answer for every failure, as for a lot of football coaches, is to work harder. Whatever we were doing before, we'll just do more of it. We'll do it longer and harder and damn it, we'll get it right.

The next day against the Amsterdam Admirals, we win the game and bring our record to 3-3.

On the bus ride back to the hotel, my friend tells me that he's really looking forward to the few days off. He has a fantastic plan. He's going to take a Viagra and masturbate all day. He's my next-door neighbor at our hotel in Hamm. I make sure not to be there: hairy palms and such, oozing through the German walls, contaminating our house of purity.

The next week we take the train to Berlin to play the Thunder. We occupy a whole car and are required to wear our team-issued maroon Rhein Fire jumpsuits during transit. After losing the game, we get back on the train to go home. We are tired and losing sucks. I take off the jacket portion of my jumpsuit and have a white T-shirt on underneath. While I'm standing in the aisle talking to a few teammates, Whiskey Pete comes walking through the cabin.

—Where's your jacket, Nate?

—It's over there on my seat.

—Go put it on. What do you think this is?

—A train?

The next day I have an envelope in my locker. I open it. It's a fine: two hundred dollars for improper road game attire. I guess "be flexible" only applies to the players.

We are back to the Relexa for the last three weeks of our journey. I catch a few passes in our next loss and am working myself back into the fold. But my hands are rusty. For four weeks in Alabama, there were no footballs around. One day after practice we make up a game—Adam, Greg, Chad, and me. There are soccer goals around the periphery of our field. One of the quarterbacks stands at the top of the penalty box and tries to throw the football past me into the net. The quarterback can work on his accuracy while throwing hard and I can work on my reaction and ball skills. Then we flip the drill and I step out to the top of the box and unleash the cannon. The weapon attached to my shoulder has sadly been relegated to recreational duty only, but every once in a while I like to light the wick and let the dragon breathe. I played the wrong position.

Back at the Relexa, I'm preparing for a visitor. Alina is coming to town for two weeks. We planned it months ago. When I first found out I was going to Germany, I was torn about how to handle it with her. We've been exclusive and devoted to each other but I have my doubts about all of it. I was shipped to Denver, then spent my first season in the NFL chained to my cell phone. And once you get started in on all the phone calls and text messages, it's hard to go back. She wields the phone like a razor blade.

When she arrives, all sixty-six inches of her toned and tanned body, her doe eyes, and her bright, dimpled smile, she spends her first few days sleeping off the jet lag. I go down to the meal room in the morning, say "Guten Morgen" to as many people as I can, eat breakfast, then bring a waffle with Nutella upstairs to my slumbering sweetheart before leaving for work. She'll fall in love with Nutella in Germany, just as I have. Above all things, Nutella is the best symbol of our relationship: sweet and delicious and since presented at dawn, seemingly nutritious, but altogether unhealthy.

Our second to last game of the season is in Cologne, a beautiful city with a cathedral near the train station. It's enormous and incredibly ornate, each inch seemingly carved with a scalpel. I stand at its base, neck craned skyward, and think about every hand that went into building it, and every heart that believed it was doing God's work. I snap some photos and keep walking.

Alina sits in the stands at the game with the other wives and girlfriends and poses for pictures, doing the cute-chick half-squat, hands on bent knees pose, with the field in the background. I play a lot and make a few improbable catches, courtesy of Greg, which make me feel like a wide receiver again. We lose, but no one really cares, myself included. It's my time now. In a few weeks I will be back on U.S. soil lining up to run routes against our newly signed cornerback Champ Bailey (I read about it on the Internet) and wearing my Broncos uniform again. As far away from that reality as I am on the steps of a European cathedral, I know that very soon none of this shit will matter. Champ won't be asking me about my time in Germany and Coach Shanahan won't be asking to see the photos of the church.

The day before she leaves, Alina makes me a cake for my birthday and buys me balloons and a pair of black high-top Converse shoes. I'm turning twenty-five in Deutschland. Somehow she makes a cake with no kitchen. She's crafty and sweet and caring. Too good for me. I'm a selfish professional football player, in constant pain and emotionally unavailable. That night a group of us dress up and go to a casino an hour away, in Dortmund. We look nice but not nice enough. They require a coat and we are coatless. We rent four of Germany's finest discarded casino jackets at the door, buy cigars, order cocktails, and set the place on fire playing blackjack the American way: loud and risky.

Germans sit quietly and brood over their cards while Adam, Greg, Bret Engemann, and I bet recklessly and smoke and drink and raise a ruckus in our house jackets. The lady to my right curses me as I chase my perfect hand. She gathers her chips and storms off. We laugh and collect our winnings.

A few days later I say goodbye to Alina and start to pack up my room. It's our last week in Düsseldorf. We are 3-6 as a team. At the end of the week we check out of the Relexa Hotel for good. I say goodbye to the friends I have made: hotel employees, cooks, Markus, stadium workers, marketing people, the Internet café guy, the gyro guy, the bratwurst guy: poof. Gone forever. I'm getting used to that.

We board a train to Amsterdam.

Adam is hurt so I get my first start of the season. It's a full German circle. All of the pain and rehabbing and traveling and practicing brings me here to Holland, running out of the tunnel as the starting receiver for the Rhein Fire in the last game of a losing season. Does it matter? Is anyone watching?

We lose the game and finish 3-7, second to last place. I have 100 yards receiving and feel solid on the field for the first time in months. After the final whistle some of us stand on the field and chat. Football players are shuffled around leagues and teams constantly. I always have a friend or two on the other team. Plus the Broncos allocated guys to other teams, too. A few of my buddies play for the Admirals. We share stories about the season. See you back in Denver! Back in the locker room, Pete puts a cap on our season.

—Guys, bring it up. Take a knee. Bring it in close. Listen guys, I know this season didn't work out how we had hoped, but I'm proud of everyone in this room. Hell, I know what kind of players you guys are. We couldn't catch a break this season but you guys never complained and you worked your tails off every day. And I appreciate all of it. Shit, sometimes hard work isn't enough. Sometimes things don't go your way. I just hope you guys enjoyed the experience as much as I did. Either way, we'll all remember this forever and I wish you guys luck in the NFL or wherever you end up. If there's anything any of us coaches can do to help, please ask us. I've got nothing but good things to say about every one of you guys. You hear me? I mean it. Now let's break it down before I lose it. C'mon, bring it in tight.

We stand up and bring it in close for a breakdown. Coach puts his hand in the air and we stack ours on top of it. Win or lose, this is the most honest moment for a football team.

—Fire on three. One! Two! Three!

—*Fire!*

Ashes.

We sit in the locker room, slowly removing our Fire gear for the last time, talking and thinking and stuck between sentimentality and a powerful relief. We shower and get in line for our exit physicals. At the front of the line I'm greeted by an old friend.

—Hey, I know you!

—Mayfield! They made you come all the way out here for this?

—Let's not talk about that, Nate. Great game. You looked good out there. How's it feelin'?

—Man, it's pretty good. I can't complain.

—You're not just sayin' that?

—No, I'm serious. It held up great.

—And what about that pinkie? Let's have a look. That thing may never be right. You okay with that?

—Yeah, I'm okay with it. I've got nein more.

—Well all right then. Good luck to you, Nate. You're gonna do just fine. I'm not worried about you at all.

—Thanks, Mayfield. You're one of the good ones.

—Don't tell anybody.

I have a clean bill of health. The journey is almost over. But first we have a night to ourselves in Amsterdam. And I want to put a cap on the season, too.

Those of us poor bastards with wives and girlfriends wander through the red-light district like gawkers at a zoo. We look into the weed cafés, the bars, and the brothels and I gauge the reactions of my friends. They want no part. So instead we go back to the hotel and stay true to our ideas of love. I watch a naked ping-pong match on television and fall asleep.

The next morning, some of my adventuresome teammates stumble onto the bus, eyes like pennies, smelling of the industries that I denied myself. I close my eyes and take a deep breath through my nose as they walk past me on the bus. That will have to do.

The bus revs up and it's time to go home. I mill around the Amsterdam airport and talk with a few players from the Admirals while we wait for our respective flights. Life is weird, we agree. It's a tough road ahead for all of us. We agree on that, too. If we were in NFL Europe at all, we are long shots in the NFL. Yet there's an unmistakable optimism. We've just been through something rarefied together. And we are all better for it. How, we don't know yet. We're tired and anxious to get home. But there's a new look in our eye that won't go away.

A football dream is easy to spot. Turn on *SportsCenter* and they'll show you what it looks like. Tom Brady's life. Peyton Manning's life. Fairy tales. Storybooks. The football dream I had as a child unfolded much differently. But it has still unfolded. Every crease and every line, ever grunt and every pop, I'm playing the game I love. The grass is still green, the hits still hurt, and the ball in flight is still the most beautiful sight I know. I will chase it to the ends of the earth.

Three flights and twenty hours later, I'm back in Denver. A Broncos staffer picks me up from the airport and takes me straight to the doctor's office for an evening physical. We have minicamp practice the next day. I pass the physical and am back in my number 14 in the morning, shaking off the jet lag and reintroducing myself to all things Broncos, no things Fire. Be flexible.

4

Grid-Irony
(2004)

It takes a village to raise a jock.

I arrive in Denver to a new group of guys. Along with Champ Bailey, we've signed another future Hall of Fame defensive back: John Lynch, fortifying an already solid defense. It doesn't take long to realize that their reputations are well deserved. Champ's calm and collected: patient and deadly. He makes it look easy on the field, never sweaty or out of breath. John's an assassin on the grass and an ambassador in the locker room. He gets along with everyone, upstairs and downstairs. Our veteran leadership is fully formed on both sides of the ball.

In addition to our new free agents, there's our new draft class.

We all sit together in the theater-style auditorium for training camp's introductory meeting. The seats are full—every one of them—with the eighty-plus players allotted for training camp rosters. Seated in the front of the room behind the podium in folding chairs are the various departments of the franchise, at least a hundred people. They squirm in their seats as we stare them down. We aren't used to having civilians in this room. They have a peculiar look to them: so small, so delicate. One by one, Coach calls up the designated spokesman for each department.

General Manager Ted Sundquist takes the podium first and introduces his scouting department. We don't see these guys often. And they don't see anything but football. They're overweight and underfucked. Sometimes they stand around and watch practice but mostly they are regional scouts scattered around the country, in charge of gathering information on college prospects. I sit in my chair and glare at one of the scouts named Bobby, hoping he'll notice me noticing him. His father is also a scout. I didn't know who Bobby was until I got back from Europe. My old Menlo coach Dave Muir is coaching at Idaho State now, and Bobby came to campus to scout one of their players for the upcoming draft. When Dave asked Bobby about me, Bobby chuckled and said I wasn't worth a shit: the number five receiver at best, probably won't make the team. Dave is feisty and very loyal. He told Bobby to go fuck himself. I knew Bobby knew I knew. I tried to get him to look in my eyes but he wouldn't. I always find it funny when a scout says a professional athlete is a piece of shit. By his own standards, he must be a Mount Kilimanjaro–size pile of shit.

The message from Ted: not only are your coaches watching every move you make, but so are we.

Greek takes the podium next and introduces his assistants: Corey and Scott. Then he gives his spiel about player health, which amounts to an emphasis on punctuality for treatment sessions. Report your injuries and treat them like a good boy. Oh, and there is a once-a-year drug test.

—If you can't pass it then you're either stupid or you have a drug problem. Either one, you need help.

Then Dr. Boublik, our team orthopedist. He looks like a doctor, slight guy with brown hair, and gives a brief overview of injury treatment protocol. Dr. Boublik has a measured, precise tone. He says that he and his partner, Dr. Schlegel, are here to provide us with "the best-quality care" and to make sure "we get healthy and get back on the field in a safe and timely manner." He speaks in the platitudes consistent with his profession, speaking to mandolescents about injuries they aren't meant to understand.

I glance around at my teammates. They all have that vacant football meeting look on their faces. They are watching the man who is speaking but they don't hear a word he says.

Our public relations director, Jim Saccomono, is next, a middle-aged man with a flair for the dramatic. The look on his face suggests that he's thought of a great joke about you but he's decided not to tell it. He introduces his assistants, Paul, Dave, and Patrick. All of them are fiercely loyal to the Broncos brand and are excellent communicators, in charge of brokering the often strained relations between the media and the players. They are also very nice people.

So is Chris Valenti, aka "Flip," our equipment manager. Managing the equipment of an NFL team is a massive undertaking. Flip has three full-time assistants: Kenny, Harry, and Jason. They are some of the hardest workers in the building. Tennis shoes, cleats, sandals, socks, shorts, girdles, jockstraps, knee pads, thigh pads, practice pants, game pants, sweats, T-shirts, practice jerseys, game jerseys, gloves, elbow pads, sweatshirts, hats, beanies, jackets, wrist bands, rib pads, shoulder pads, neck rolls, chin straps, mouthpieces, face masks, and the tip of the spear: the helmet. Not only do they give us this equipment, they also keep it clean and shiny. Every single tool of the trade goes through Flip and his boys. Even sunflower seeds. They're good friends to have.

Then Ken, the video director, takes the podium. Every practice is filmed from two angles: the sideline and the end zone. The two angles are spliced together back-to-back. With the advancement of computer technology, the viewing of film has become very specialized. Ken and his staff are at the forefront. Any game played by any team in any year can be pulled up in an instant, with ungodly permutations of specific plays.

—Hey, Ken, could you make me a DVD with every one of Randy Moss's red-zone pass plays from week four to week eight of 2001?

—Sure thing, Nate. Come by later today and we'll have that for you.

The only thing we can't do is take home DVDs of our own practices: just in case they fall into the wrong hands.

Then our head strength coach, Rich Tuten, who strongly resembles Sergeant Slaughter without the hat. Rich stands barrel-chested at the podium and unfolds a piece of paper, eliciting snickers from the gallery. He introduces Greg Saporta, his assistant. Greg's also known as "Crime," as in McGruff the Crime Dog, for his aggressively strict eye on the off-season conditioning program. Crime's head is freshly shaved with a straight razor. He smiles only if he absolutely has to. Then Rich lists the players who have earned a perfect attendance record during the off-season conditioning program. Rod's name is read for the twelfth straight season. He's Rich's model student—he wasn't drafted coming out of college, and he spent his first year on the practice squad. Bobby must have thought he was a piece of shit.

Then Coach introduces local TV personality Reggie Rivers. Reggie was a Broncos running back in the 1990s who now works for CBS, the main Broncos affiliate. Reggie takes the podium and sums up the local media. The *Rocky Mountain News* and the *Denver Post* are competing newspapers. He rattles off the names of the competing television and radio stations. All of them, he says, are running the same stories and trying to find ways to set themselves apart. His presentation amounts to a list of dos and don'ts when dealing with these hacks. He cues up a film and shows us examples. Do say: We're taking this thing one game at a time and we'll see what happens. Don't say: Man, I really would like to go home and eat a heroin sandwich.

The next morning we're back into the crush of training camp and Champ's giving us fits. With pads on, his physical dominance is on full display. He's an excellent cover corner. He's in his physical prime. He's smart and tough. But where he excels the most is on the line of scrimmage. We can't get off his press. He jams us at the line as soon as the ball is snapped. But I love it, because I know that there's no one in the world better than Champ. Hence the name.

I'm in good football shape from being in Germany. My body feels fine. My knee's healed. My pinkie's crooked, but oh well. The only thing bothering me, except for the typical early training camp agony, are the terrible blisters that Champ is indirectly causing to bubble on my feet. All of the friction and combustive energy I push through my body comes out through the balls of my feet and my big toe. After the first day of camp, my feet are hamburger meat. They get worse and worse with every practice, until the pads are down to the bright red flesh. Before every practice I jump up on the training table and our trainer Corey jellies, gauzes, and wraps my feet the best he can. It doesn't do much. Every step feels like I'm walking on hot coals. By the time practice ends, the toes of my socks are soaked red. Every night, I limp into the meeting room and fall back into my chair. We sit in a cluster near Blade on the right side of the room.

An average night of film goes like this:

—Now look at Rod here, you guys. See how he sets him up?

Rod has a head-up corner and is split wide to the right. His route's a slant. He gives a quick head fake off the ball, pushes up the field three hard steps, stops on a dime, and undercuts Champ, swimming him by with his left arm and breaking toward the middle of the field. Pass complete.

Slo-mo, rewind, slo-mo, rewind, slo-mo, rewind: ad infinitum as Gary Kubiak, our offensive coordinator, talks about the play. Kube, as we call him, is an easygoing Texas native and a former NFL quarterback who played behind John Elway for his entire career. He's already won three Super Bowls as a coach. He finds success wherever he goes, and he runs our offense day to day.

—Now, if he's playing head-up on you like this you have to get him to move his feet or you've got no chance. Nice work, Rod. See how patient he is, guys? See how he makes him think he's going deep? You've got to sell it, boys, especially against Champ. He's seen all the tricks in the book. Great job, Rod.

The next play is a running play. Charlie and I are on the field. It's to my side: 18 toss. I block the corner. The safety to my side makes the tackle.

—Great hustle on the backside, Charlie. Nate, who are you supposed to block on this play?

—Depends on the coverage.

—What coverage are they in here?

—Three.

—Yes, but it's a Three Cloud: we call that Four.

—I thought Four was quarters.

—Some places it is but we call quarters Cover Eight. You should know that by now.

I clear my throat, say nothing.

—So if it were a regular Three on this play then the safety to your side would be deep middle and the strong safety would be down near the box over here.

He circles an area of the big screen with a laser pointer.

—Then you would be right, it would be Three, and you would block the corner just like you did. But that safety is down to your side. They rotated weak. It's a Cover Four. So now what's your rule?

—Push/crack.

—That's right, push/crack. Push up the field at the cornerback then crack down on the strong safety as he comes up to fill the run. If he never comes up to stop the run or if the corner gets horny, you stay on the corner. But this crack-back can be a kill shot, boys, and the difference between a three-yard gain and a touchdown.

—What if they're in Cover Four, er, Eight?

—Then you'll stay on the corner, unless that safety is really cheating or shows blitz late and we don't have time to check out of it. Then you *have* to get him or the play is toast. Same with Cover Two; keep an eye on that safety. But be careful because that corner has run responsibilities in Cover Two so if you leave him for the safety the corner might just come up and make the tackle. What you'll have to watch for is that Cover Six: the quarter-quarter-half. That's Cover Two on half the field and Quarters on the other half. But it really doesn't make a shit what they do. Just watch that safety to your side and if he gets horny you block him. If not, stay on the corner.

—What about man coverage? Should I run him off or block him?

—If you know he's in man you can outside release and run his ass down the sideline. But if he smells run then you gotta block his ass. Remember, if you guys can't block, you won't be on the field.

Rod chimed in.

—You know how to tell if they're in man, right?

—Uh . . .

—First off it's in their eyes. If they ain't lookin' at nobody but you, it's probably man: especially if they're up in your face. If it's zone they'll be looking in at the quarterback and looking inside at the other receivers. But be careful, sometimes they'll be right up in your face breathing on you and you think it's man, then they bail out right before the snap and drop into their coverage. You just gotta feel that.

—All right.

—But look, you see that tight end motion to the other side of the ball?

—Yeah.

—What happened when he motioned? What did the defense do?

—Nothing.

—Exactly. So what does that tell you?

—It's zone.

—Yep. If it was man, whoever was covering him would've run with him, 'cause he'd have him in man coverage. But since no one moves, you know it's zone.

—Exactly. That's why we do those shifts and motions, boys. It ain't just for the hell of it, all right? We're trying to give you guys a pre-snap read so you know what's coming. We're trying to gain any advantage we can. Every little bit helps. Okay, next play.

On the movie screen, Ashley Lelie runs a comeback; twenty yards straight up the field and back downhill at a sharp angle toward the sidelines. Champ undercuts his route and bats down the ball like he was knocking an ice-cream cone out of Ashley's hand.

Blade leans over to Ashley and whispers.

—On this comeback route, you're really going to want to come *back*.

Blade's dealing with what all former-players-turned-coaches deal with: explaining in words what they always knew how to do instinctively. We all know what Blade means. But Kube's coach-speak is more refined. He takes it a few steps further.

—Ashley, you gotta come off the ball harder and *attack* his leverage. And get your full twenty-yard depth. Not eighteen, not nineteen—twenty. And keep your shoulders over your toes on your break. Don't stand straight up and chop your feet at the top of your route like this. See, that's when Champ reacted to your break, right when you started chopping your feet. And don't veer off that straight line. See how you are veering off right here?

Laser pointer.

—No way that'll work. Look at that. When you veer off like that, he knows you're not coming inside anymore. Look at that. He read your ass like a book. And you gotta stick that inside foot in the ground at the top. Stick it in the ground!

Slo-mo.

—And keep those elbows high and tight and them arms pumping. Pull that elbow through, get your fucking head around, and come *downhill*. You got me? Come downhill at a sharp angle and *attack* the football. C'mon, Ash. We gotta get better at this.

It is a difference in schools of thought. One school, Blade's school, trusts the instincts of the pro football player, because he trusted his own when he played. He allows more improvisation. He supplies the general parameters and steps back. The nuance of the game's technique is decided by the player's athletic instinct.

The other school, the one more common in the NFL, is the more rigid, systematic assembly line of angularly identical patterns. It believes that every football play has one right answer. If you choose the question you get to choose the answer. It is a tightly structured philosophy and has evolved steadily over the years. Blade and Kube both played in the NFL, but they had very different experiences. Blade was a wide receiver and played in every game. Kube was a career backup quarterback who knew the system inside and out but rarely got to play. With all of that studying and no playing, the game becomes conceptual, and as a coach, Kube trusts the concepts over the instinct of the player, who comes and goes. But Kube skillfully toes the line between player-speak and coach-speak, and knows how to communicate in terms we understand. And he serves as the perfect buffer between the rigid offensive system and our often unsystematic instincts as players.

They're always hard on Ashley because he's a first-round pick. They expect a lot out of him. They also expect a lot out of Darius. He's a rookie second-round pick. And he's up next. He runs a skinny post versus off coverage. The ball's slightly behind him. Incomplete pass.

—Okay, D. You know what I'm going to tell you here, right?

—Catch the ball?

—No. Well, yes, but no. How many steps do we take on a skinny post?

—Seven.

—Right, seven. Now let's watch your steps.

Slo-mo.

—How many did you take?

—Five.

—Yep. That's why the ball was late and behind you. That quarterback is taking his drop and letting it fly on time. Look, you broke at nine yards. This is a twelve-yard route. Remember, guys, there's a rhyme and reason for everything. The skinny post is a seven-step route, ideal against Cover Three or Quarters, but it takes timing. First off, your inside foot has to be forward. Make sure your inside foot is up or else you're gonna be breakin' on your sixth or eighth step, and the timing'll be thrown off. It's seven steps. So your inside foot is up, seven steps hard up the field and break to the near goalpost. And don't take yourself across the field. Remember, it's called "skinny" for a reason. We are staying in that seam. Right up the sidewalk. Got me, D?

—Yes, sir.

—And you're right, catch the fucking ball.

Next up is Charlie. He runs a "sluggo": a slant and go. The corner bites on his slant fake and Charlie runs right by him. Pass complete for a touchdown.

—Charlie, great work here. You guys see this? This is textbook stuff right here. What's the most important thing on the sluggo, guys?

—Sell the slant.

—That's right. You gotta sell the shit out of it. And that doesn't mean a half-assed head fake. Run the slant for three good steps. Make him believe you. Give the quarterback your eyes. Then come out of it hard and give your quarterback a throw. Be a friendly target for him. You have time to win on this route. We are going to protect it up front, but you gotta beat your man. This is a home run if we do it right. Great work, Charlie.

Then our other rookie, Triandos Luke, is on-screen running an "acute": a twelve-yard comeback. Champ jams Tri off the line and lets him go, indicating that he's in a Cover Two. Tri runs his twelve-yard comeback but the quarterback, seeing Champ sitting in the throwing lane, goes elsewhere with the ball.

—Tri, you know what I'm going to tell you here, son?

—No, sir.

—What coverage do you see here?

—Cover Two?

—Yes, Cover Two. And what do we do on the Acute versus a Cover Two?

Rod leans over and whispers to Tri. Tri relays it to Kube.

—Convert it.

—That's right, you gotta convert this to a fade. There's a hole shot there if Jake wants to take it but with you out there running your own shit, we don't have a chance. See how Champ is sitting in that window? This ain't a Cover Two route. We gotta have our heads up and be payin' attention out there. Don't get locked in, boys. Every play depends on what they give us. We gotta know that we can count on you guys to know your shit. You got me?

—Yes, sir.

—Okay, fellas, that's enough for tonight.

It was a two-hour meeting to end a thirteen-hour day.

—Good job out there today. Look, I know y'all are tired. This shit ain't easy. If it was easy, everyone would be doing it. We're going to be hard on you in here so you can improve as players out there on the field. You got me? Just keep on fighting, one day at a time. Let's get a little better every day. Break it up with your coaches.

We stay in the room with Blade, and the QBs, running backs, and tight ends leave to their respective position rooms. Every position group has its own meeting room. We have fifteen minutes to kill before he can let us go. At 9:30 p.m. our day's over. I stop by the cafeteria on my way to my car and fill up a Styrofoam to-go box with whatever late-night snack options I can find: corn dogs, taquitos, burritos. I drive home in pain and silence: another day of training camp in the books.

The first ten days of training camp are the roughest. Each day feels like a week. Each step feels like my last. The pain is constant and comes from everywhere, pushing up from the bottom of my feet and down from the top of my rattled skull. But soon there is a preseason game that breaks up the monotony of the ritual lashings. Then after the second preseason game, we prepare for the first of two rounds of cuts. One is after the third preseason game and one is after the fourth and final. The first cut chops off about twenty guys. I look around the meeting room after the first cut and feel the weight of their absence, knowing damn well I've just witnessed twenty football funerals. Flip makes it official as he goes from locker to locker removing their nameplates and filling up his rolling bin with helmets and shoulder pads.

The night before the last preseason game that will set the final roster, Kube excuses the starters from the meeting room and lets them go home and rest. They won't be playing in this game. They've already made the team. Must be nice. The rest of us are just hoping for another few weeks of free shoes.

—Now I know this isn't where you wanna be sitting, boys. You wanna be out there with those guys. But don't get discouraged. We brought y'all here for a reason, and you guys played your asses off for us all summer. Don't think we didn't notice that. There are some great football players in this room. So tomorrow night, when you get out on that field, leave it all out there. Think of it like this: it's a job interview for thirty-one other companies. All of them will be watching. If it doesn't work out here, another one of those thirty-one teams might snatch you up. Put it on tape. Put it on tape. The eye in the sky don't lie, boys. That tape lasts forever. Be proud of what you've done here, but finish it off the right way tomorrow. Believe me when I say it, I wish we could keep all of you guys. I really do. But we all understand this business. This is the hardest part of the job.

A lot of coach-speak is bullshit, but I believe Kube when he says this.

—I'll leave you with this: whether or not this is the last game you ever play, understand that by making it into this seat at all, you have beaten the odds. You should feel proud of yourselves for that. Now go out there tomorrow night and have some fun. Pin your ears back and have some fucking *fun* out there, boys. Good luck.

• • •

The next night I run out of the tunnel and look up in the stands. My mom and dad are sitting a few rows back with Alina. They wave down at me as I jog to the sideline. Coach said we'd get plenty of playing time in the game. That's good and bad. Football game shape is different than football practice shape. And I'm on every special teams unit. I brace myself for a long evening.

We rest our starters but Arizona plays theirs in the first half. Matt Mauck and Bradlee Van Pelt have been locked in a solid backup QB battle all camp. They are very different quarterbacks, trying their best to adapt to the demands of a new system. The over-coaching and attention to detail that NFL quarterbacks must deal with is stifling. For a rookie QB, it can be overwhelming. But Bradlee and Matt have made it through the hardest part. Now it's time to play. Matt starts the game and takes the bulk of the snaps. Charlie, Triandos, and I play the whole game at wide receiver, rotating as much as we can but often stepping to the line of scrimmage horribly out of breath. In the middle of the third quarter I run a backside post on a pass play designed to go to the other side. The backside post never gets the ball in practice, but shit happens during games. Matt scrambles and finds me running across the field. He throws a low strike across his body and I go down to the ground to secure the catch. Touchdown. Triandos smacks me on the head. I hold the ball up high in the air. Matt and I share a quick embrace on the sideline and I'm back in the huddle for kickoff.

We lose the game but who cares. I played well. Charlie played well. Triandos played well. Matt played well. Bradlee played well. But some of us will have to go. Vic Lombardi, CBS's main sports anchor and the face of Denver sports media, interviews me after the game.

—How do you feel you did out there tonight?

—I feel pretty good about it. There's nothing I can do now, though. Coach said to be near our phones tomorrow morning. That's when they're doing the cutting.

—So you're saying I shouldn't call you tomorrow morning then?

—No. Please don't.

He turns to the camera.

—You heard it. If you know Nate Jackson, do *not* call him tomorrow. From the Broncos locker room, I'm Vic Lombardi, CBS Sports.

The next morning my phone doesn't ring. I check it to make sure the battery isn't dead. The following day I tiptoe into work and look around. My locker's still here. No one is looking at me funny. I made the fucking team. Triandos is here, too. So is Charlie. So is Matt. So is Bradlee. We all made it, for now.

Coach Shanahan takes the podium at the first team meeting of the regular season.

—Look around the room, guys. This is our team. Looks pretty empty in here, doesn't it? Well, you guys made it. And you all deserve to be here. You've earned that seat. And I couldn't feel any better about the guys in this room. We have a chance to do something special here. Now it's up to us to see if we can put it all together. We know what we've got to do; let's go get it done.

That night I talk to Ryan, my agent. He congratulates me and says this is just the next step.

—Keep doing whatever they ask you to do. Get in tight with the special teams coach. The more you can do, the better.

We also talk about my contract, but there's not much to discuss: it's minimum wage. There's also a "split" clause in the contract stipulating that if I get put on injured reserve, my pay gets cut in half for the remainder of the season. Much like the waiver I signed in San Francisco, the split is a product of having no leverage. Whatever. I'm on a team. That's all that matters. And teams play games. But games only take up sixteen days of the year.

The other 349, I'm still a professional football player. Those days don't make it on television, but they are what the job is all about.

With all of that time around the games, football becomes a science project in technical perfection. That means execution. The NFL is about execution.

But it's not hard to figure out how to do things right. I just watch Rod. He'll let me know when I'm messing up. He cares too much about the team not to. That week we win our home opener against the Chiefs. Then we travel to Florida the next week to play the Jacksonville Jaguars. I'm the number four receiver. My job's to give Rod, Ashley, or Darius a rest if they get tired. Sometimes I'll stand at attention for a whole quarter and not get on the field; others I'll have a flurry of plays.

In the locker room at halftime, Jake's pissed. We're down 7–6. The offense isn't clicking. The receivers aren't producing and look tired on the field.

—We need someone in there who can fucking run! We need some fresh legs! Fucking put Nate in! At least he'll *run*!

Late in the fourth quarter D-Watts is slow getting up after he fails to come down with a ball on the sideline. Blade looks at me.

—Get in for Darius!

I flag him down and run on the field. There are just over two minutes left in the game and we're on our own thirty-two, down a point. I run an "Omaha"—a five-yard out—and Jake throws me the ball. I catch it and am tackled immediately as the clock strikes the two-minute warning. My first NFL catch. They honor the moment by going to a commercial break.

Coach motions for me to stay in the huddle. We're in our "zebra" package—three receivers, one tight end, and one running back. Our first-down pass falls incomplete to Rod. On second down, Jake hits Ashley for eight yards. The clock ticks. We run up to the line of scrimmage. I'm in the slot. Jake yells out the play and snaps the ball. I run 10 yards straight up the seam and bend it in toward the middle to see the ball spiraling toward me. It sinks into my chest as the free safety pops me on the left side of my body. I fall to the ground with the ball in my arms. First down at the Jaguars' thirty-five: field goal range for our Pro Bowl kicker, Jason Elam.

We call timeout. I jog off the field, replaced by a running back in our base package. My teammates high-five me and slap my head.

On the next play Jake runs for nine yards, then Quentin Griffin for two yards and a first down. Jake spikes the ball with 1:07 left. The next play's a one-yard gain down to the twenty-four. We just have to center it up for Jason. But on the next play—with forty-four seconds remaining—Q takes the handoff and fumbles. Jacksonville recovers the fumble and we lose, just like that. We hang our heads and walk across the field to the locker room. Blade puts his arm around me on the way in.

—Nice catch, Nate.

I shower and get on the bus and eat a turkey sandwich from a boxed lunch that was in a bin outside the locker room. It also has a Snickers in it. And some potato chips. And an apple.

After that game we rattle off four straight wins. One of them is a trip to Tampa to play the Bucs, John Lynch's former team. He had played his entire career there up until that season. The crowd gives him a standing ovation when he's introduced. They shoot off the cannons from the pirate ship in the north end zone. He waves to the crowd. He's a legend.

Darius, meanwhile, is back in good standing with the coaches so I play mostly special teams. That's how the season goes: a few plays here and there on offense and a whole lot of special teams. On Wednesday morning of every week I sit down in my seat at 8:15 a.m. Special teams coach Ronnie Bradford takes the floor with an overflowing cup of coffee and presents the English language in new and exciting ways. Meetings are only as entertaining as the coach who is conducting them. In this case, they are endless fun. Ronnie says the words *gentlemen* and *okay* so much that we have a daily tally going. He once hit seventy-seven *okay*s in a thirty-minute meeting.

I am learning the mental side of the NFL, mastering the daily rigmarole that it takes to be a pro. It's not just out there on the field. That's just a fraction of it. It's in here, sitting in this cushy leather chair, watching film and listening to coaches talk for three or four hours every day. This is the hard part: figuring out what to think about. The mind steers the body, and my body is developing a new problem. My aching Achilles has prompted a reevaluation of my shoes and insoles. Maybe I need lifts in the shoes to keep my heel up higher. And some soft insoles that would conform to my feet and supply better support. I try the new foot setup and sure enough it alleviates some of the Achilles pain.

But soon the bottom of my foot begins to hurt. The connective tissue between the heel and the ball of my foot, the fascia, becomes tender and painful. It's plantar fasciitis. Plantar fasciitis is similar to the Achilles pain in that it's chronic, nothing gives it relief, and it hurts with every step. But after thirty minutes of practice it warms up well enough to get through the day. Yet no treatment helps: massage, ice, ultrasound, stim, stretching, meds, acupuncture—nothing. They decide to get an MRI.

The MRI results from the lab:

IMPRESSION:

1. Focal moderately severe acute on [*sic*] chronic plantar fasciitis of the medial chord of the plantar fascia, just distal to the calcaneal origin with a suspected component of chronic partial tearing but no fluid filled defect. There are surrounding associated soft tissue inflammatory changes.

2. Inflammation in the sinus tarsi.

3. Very mild posterior tibialis and peroneal tenosynovitis.

4. Vague bone edema in the cuboid and to a lesser degree in the anterior body of the calcaneus. There is no stress fracture and no evidence of acute trauma. The findings may

be related to a chronic stress response or nearly resolved contusions.

Boublik's assessment:

Nate comes in re-examination of his left foot. He states that his foot feels about the same. He continues to have some soreness along the medial proximal plantar aspect of the foot.

Left foot persistent symptoms in the region of the proximal medial plantar fascia. The patient is not improving with conservative treatment. We will obtain an MRI of his left foot to fully delineate the extent of his injury. Follow-up discussion of definitive treatment after MRI available.

ADDENDUM: I spoke with Dr. Crain by telephone today regarding Nate's left foot MRI. Her verbal interpretation was plantar fasciitis with involvement primarily of the medial cord. . . . There is no defect, but there is a component of partial tearing. There is also some mild first MTP degenerative joint disease and mild sinus tarsi inflammation and mild cuboid edema.

I had already spoken earlier with the player regarding the possible role of a plantar fascia injection. We discussed the risks including the fact that the injection is somewhat painful, risk of infection, risk of rupture of the plantar fascia, and the risk that he may not get complete relief of his symptoms.

I shrug off the risks and accept the offer to inject, thus beginning my long relationship with the needle-as-savior approach to injury treatment. Toradol, Bextra, Kenalog, dexamethasone, Medrol, cortisone, Ketrax, PRP; the needle is the last resort when the pain is too much or progress is too slow.

Accompanying the injection is a new type of shoe insole: a hard, plastered orthotic. Greek imprinted my feet and sent away for it. A week later, orthotics specifically made for my feet arrived in the mail and I am told to use them in my shoes. They put my feet at an angle that science apparently finds pleasing but I find compromising. It feels like I'm standing on a pile of rocks.

—Man, Greek, these things don't feel good.

—It takes a while to get used to them, Nate. Just give it a little time.

A few weeks later we play the Chargers in San Diego. The orthotics fill up my shoes and my feet feel like cinder blocks. On a ball intended for me in the end zone, I get tangled up with a cornerback. He falls sideways on one of my cinder blocks and I feel a crunch in my ankle. I try to walk it off. I don't know it yet but my tibia is broken; a vertical fracture through the bottom bulbous part of the bone: a curious little break, everyone will agree.

After the game I limp onto the bus and call Alina, who's staying at my apartment in Denver. Our long-distance relationship is coming to an end. She's just been accepted to Denver University and has decided to transfer schools and move here. Her real-estate-savvy parents bought her a loft downtown. It's official. They had just signed the paperwork that day and flown back to California. But Alina stayed at my place so we could have a few days together before she went back to Santa Barbara to finish her semester.

While I was breaking my leg in the north end zone of [Insert Corporate Logo Here] Stadium in San Diego, she was skimming through some files on my laptop. She came across a most unfortunate confessional I had written more than a year prior that outlined in detail a regrettable rendezvous I'd had with my downstairs neighbor shortly after moving to Denver. The night it happened, Alina was visited in her sleep by the phantom of my deviance. She called me in the morning and asked me if the phantom was real. No! I assured her, and rushed to my keyboard to repent, if only to myself.

When I call her from the bus she's hysterical. She explains what she has read. I'm too weak to defend myself. I admit to the transgression, hang up, and sigh. My leg's the least of my worries. I have four or five hours to prepare a concession speech. By the time the plane lands I can't walk. I hop all the way to my front door on one leg. Sweating and whimpering, I swing open the door to find her sitting on the couch like a powder keg, ready to blow.

—C'mon, babe. I'm really hurting. Can we talk about this tomorrow?

—No we most certainly cannot.

Kaboom.

The next morning I'm happy to get out of the house for my MRI. Results from the lab:

IMPRESSION:

Vertical to oblique fracture involving the distal media tibia extending from the metaphysis to the medial tibial plafond articular surface/base of the malleolus, with no displacement.

Old high grade sprains/tears of the lateral and deltoid ligaments but no acute ligament injury.

Moderate tibiotalar joint effusion, and two smaller posterior loose bodies and one small anterior loose body.

Mild degenerative changes of the tibiotalar joint with mild cartilage thinning, lateral talar dome, and osseous spurring, anteromedial talus.

Mild tenosynovitis of the posterior tibial tendon with prominent surrounding scar tissue.

Boublik's transcription:

IMPRESSION: Left ankle non/minimally displaced vertical medial malleolar fracture with evidence of chronic ankle sprains and possible intra-articular loose bodies. No evidence of osteochondral injury.

PLAN: Natural history and operative and non-operative treatment for this type of injury were reviewed with the player today. This would be a season ending injury for the player whether it is treated operatively or non-operatively. The advantage of operative treatment would be to stabilize the fracture and allow more rapid rehabilitation. If we were to treat the patient surgically, then this would also give us a chance to arthroscope his ankle and potentially remove the loose bodies, if in fact they are loose. Prior to this injury of 12/5/2004, the player did not give a history of locking or catching or had a clinical examination that would be consistent with a loose body. Therefore, these loose bodies may be incidental and asymptomatic. There is always a chance of these becoming symptomatic in the future. The other treatment possibility would be to treat the patient conservatively with a boot, non-weight bearing, icing, elevation and repeat serial x-rays. The advantage of this would be that the patient would obviously avoid the risks of surgery. We would need to follow his x-rays closely for any evidence of displacement. If he were to displace, then our recommendation would change more strongly toward operative fixation. Patient is interested in being treated conservatively at present as above. We will clinically follow him closely as well with serial x-rays.

The split in my contract kicks in and I finish the season with my last four checks cut in half. I gimp around the facility like a ghost while the rest of my teammates make a playoff run. We sneak by the Colts late in the season and set up a rematch of the previous year's wild-card game, in which we'd been disemboweled. Same result this time around. Peyton's just too good.

5

Meat Sacks
(2005)

My shoes and orthotics go in the bin underneath my locker and I finish the last four games on injured reserve, and on crutches. A few months into my first real off-season, my rehab complete, Coach Shanahan calls me with a proposition:

—How would you feel about playing tight end?

He's done this before. Our offense uses the tight end beautifully in the passing game, and having a skilled pass-catcher at the position puts defenses in a bind. Shannon Sharpe came into the league as a receiver and Coach turned him into a tight end: same with Byron Chamberlain and Jeb Putzier. Shannon and Byron are gone now but Jeb is still here. He's our main receiving tight end. All I want is to run routes and catch passes. But I'm beginning to understand what NFL longevity requires: an ability to adapt.

—Sure, Coach. Absolutely.

I start eating like a fat man. When the off-season program starts a few weeks later, I lift heavier and harder. After workouts, I take Myoplex shakes home from the facility and refrigerate them. Every few hours I pour one in the blender and season it with a few eggs, ice cream, and chocolate sauce. The weight comes quickly.

So do the bowel movements. I consider starting a band called "Two-Poop Morning," decide otherwise. I go from 215 pounds to over 245 pounds in a few short months with Rich and Crime. By minicamps I've taken on my new tight end body.

With the new body comes a new job description, and a new number, 89, the only number in the eighties that's still available. Last year's 89, Dwayne Carswell, a beloved longtime Bronco, is gone now. His nickname was "House." Now that I've taken over his number, the guys are calling me "White House." I approve this message.

The tight end position utilizes an entirely different set of techniques and terminology. I never realized it before, but tight ends need to know everything about the offense: pass plays, run plays, pass protections, offensive line calls, etc. Plus I have a new group of buddies and a new position coach. My days as a receiver are over. I shed a single tear. It evaporates immediately.

My new tight end coach is Tim Brewster. He had a standout college career at Illinois in the early 1980s and a cup of coffee with the Giants and the Eagles in 1984 and '85. He's a hard-nosed, tough-talking coach who keeps a dip in his lip and a scowl on his face; ever ready to drown out the sound on the field with a "What the fuck!" for whoever messes up. It's his natural reaction to a mistake: nothing malicious, just his coaching style. But compared to Blade's calm demeanor, Brew's approach comes as a shock. Thankfully, he spares me the vitriol and focuses his attention on our newly acquired tight end Wesley Duke.

Wes was a college basketball player at Mercer University with freakish strength and athleticism but zero experience on the football field. They want to mold him into another Antonio Gates, the athletic model of the twenty-first-century tight end, one who can blow by linebackers and safeties and attack the football at its highest point. Brew was Gates's coach in San Diego and helped turn him from a college basketball player to an NFL beast. Brew figures he'll do the same with Wes. But it's not an easy transition to go from the streets to the NFL, where technique is so maniacally stressed and the terminology is a foreign language. It takes years to train your body to master the tiniest movements. Still, Wes's struggles infuriate Brew. There are five of us in the meeting room, but it may as well be just Brew and Wes.

—Wes, look at yourself here. First of all, look at your stance. If a stiff wind blew through here you'd be on your ass. You hear me? Widen your feet, Wes. And flatten your back. You gotta be in an athletic stance and ready to fucking pounce on someone. You hear me?

—Yes, sir.

—Second of all, look at this route. You look like you're out for a little fucking stroll in the park, Wesley. Come off the ball violently and run your fucking route with a sense of urgency. Jesus, you're going to give me a goddamn heart attack, Wesley. Do you even know what coverage they're in?

Brew spends all of his energy on Wes, which allows me to work on my craft in relative peace. I have already learned a crucial lesson in the NFL: watch someone who knows what they're doing and copy him. If you get tied up with nonsensical coaching tips then you get confused and you play like shit. I watched Rod for two years and tried to emulate him. Now that I'm a tight end, I tune out Brew and watch our newly acquired tight end Stephen Alexander, aka S.A. S.A. is from Chickasha, Oklahoma, and has the easy demeanor of a man who feels good in his skin. Six foot four, 250 pounds, he's in his eighth year in the league and knows all the tricks. He carries himself smoothly and calmly, which in turn soothes my nerves. Technical improvement is a one-day-at-a-time thing, never mind the cliché. And run blocking relies on a technical checklist: stance, first step, head and hand placement, hips, feet width, etc. And in the passing game, a tight end must release through a herd of meat sacks, a process of "dipping and ripping" with a checklist of its own. S.A. will show me the way. Brew is on mute.

Minicamp gives me a soft opening for my new position. The running game is toned down because we don't have any pads on and I get to use my receiver skills in the passing game. Running routes against a linebacker or a safety is easier than what I had been doing, locking up with Champ twenty times a day. But once we get to training camp and put on pads, I'm in for a rude awakening. I have to block behemoth defensive ends like Trevor Pryce, Courtney Brown, and Ebenezer Ekuban—guys who are roughly my height plus another fifty pounds. My technical improvement isn't coming along fast enough to spare the daily ass beating they are handing me. Every day when I step onto the practice field, I have an "oh shit" moment when I look up to find that my blocking assignment is the size of my old Honda. I take a deep breath and swallow it. No time to be scared. Over and over, I throw my head into the car crash.

The violent intensity of practice obscures my surroundings. But as I stand drinking water one mid-training-camp morning, a rare moment of repose, I watch our new number 19 take a slant to the house as the crowd screams their approval. My childhood hero, forty-two-year-old Jerry Rice, is my new teammate. His rookie year was twenty-one years earlier. I was six years old and wearing my plastic Huff 49ers gear around my neighborhood. By the time I was ten, Jerry was a god and I was his disciple. Posters of him wallpapered my room. "Goldfingers" was my favorite. In it he stood on a runway in a black tuxedo with a gold bow tie holding up a football. His hands were painted gold. Behind him, a private jet was parked next to a red Ferrari: the spoils of his golden fingers. My friends and I played game after game in the street. When I scored a touchdown, my squeaky voice bounced off the garage doors and echoed through the trees. "Touchdown, 49ers!"

Like three years earlier at the 49ers' facility, I'm caught now between a reverence for the legends who shaped my childhood dreams and an acceptance of my fate as a professional football player. What makes Jerry's presence even harder to reckon with is that he's on the decline. He played sixteen years in San Francisco and the previous three and a half before this in Oakland. Halfway through the 2004 season he requested a trade and went to Seattle. After the season, he signed with us. It's his twenty-first NFL training camp.

Football is a young man's game. Thirty is old in the NFL. Forty is unheard of. I marvel at his energy and work ethic. He comes out to practice early to work on footwork and stays late to catch balls. I rejoice inside every time he catches a pass. And I cringe every time Champ locks him down in man coverage. There are no fond farewells in the NFL. It is eat or be eaten: even if you're dining on G.O.A.T.

I tell Jerry once and once only that I was a big fan growing up, but then I let him be. In the locker room we are all the same. You earn your keep or you get the fuck out. The best way for me to honor my hero is to follow his example on the field and leave him alone off of it. He is the greatest player to ever play the game of football, and he is confronting some real shit: this thing is almost over.

I'm dealing with a similar fear. My new body is struggling to keep pace with the demands of camp. During a particularly exhausting practice, I run a deep crossing route in the end zone and as I try to accelerate past one of our linebackers, he tugs my jersey, which yanks my hamstring. I feel it up high, right against my butt bone, and very deep.

The rehab is ice and electric stimulation, strengthening, stretching, and cardio right away. But after three or four days, the benchmark for noticing improvement, I notice none. In the morning, I limp into the training room to see Greek.

—Feeling better today?

—Uh, maybe a little but not really.

I say "a little" no matter what.

—It should be improving by now.

He has plugged me into his how-to-fix-a-broken-jock mainframe and is following the protocol step by step.

Twice a day I go through an extensive exercise program that attacks the injury. Lots of guys get hurt in training camp. Some guys get comfortable not practicing so they're pushed back on the field.

That's done with verbal pressure in the training room, creating a timeline for return, and kicking his ass in the weight room, twice a day, so that an injured player would prefer to practice hurt than receive this isolated attention.

The pressure works, and I get back on the field in time to play the Houston Texans in a preseason game. I have to prove I can hold my own in the trenches if I want to stick around. But during the game I reaggravate the injury. When we get back to Denver I get an MRI.

Lab results:

FINDINGS: There is a moderate strain of the proximal hamstring tendon complex with more focal partial tearing at the origin of the biceps femoris and semi tendinosis conjoined complex. There is associated surrounding soft-tissue edema and hemorrhage. No underlying bone edema. The majority of the hamstring complex at the ischial tuberosity is intact. There is also interstitial degeneration or partial tearing in the semimembranosus tendon just distal to the origin.

Translation to me: mild hamstring strain. I'm prescribed exercises, ice, meds, and modalities. We good, Nate? Yeah, I guess we're good. But another week goes by and I make no progress. Greek lets me know that my progress isn't jibing with the timeline that fits the protocol for "mild hamstring strain." That gets me thinking, what's wrong with me? Am I being a pussy, or what? Greek says I should be ready to go, so I know that's what he's telling Coach every day when he gives him the injury report.

So I get on the field again in time for the last preseason game in Arizona. I'm still thinking I need to prove myself a warrior to make the team. Brew tells me he wants to see me go in there and bite someone in the fucking neck. Before the game I use heat, pain pills, meditation, stretching, Icy Hot, back plaster (a patch that goes on the skin and heats up as your body warms it), and my own natural adrenaline. None of it works. I limp around and play badly.

Oh, and I reinjure my hamstring. I think I'm toast, for sure. But my phone doesn't ring the next morning. Despite my poor training camp, I've made the team again. On the first day of the regular season they decide to ramp up the treatment plan.

Robert Williams, M.D.:

CHIEF COMPLAINT/HISTORY OF PRESENT ILLNESS: Nate is being seen for a proximal right hamstring strain. He tried to play on Friday and reinjured the hamstring. He does not have constant pain but feels as if the injury is similar to how his first injury felt approximately a week and a half after that injury. He does have some pain both proximally in the area of his origin of his hamstring, as well as some pain that he describes that radiates medially into his inner thigh.

PLAN: The treatment options were discussed with Nate of conservative management of continued stretching and strengthening and rest. It was recommended that an injection into the origin of the hamstring tendon would help control the inflammation and allow for quicker recovery. Nate was explained the risks and benefits of the injection. He understood the risks were infection or reaction to the medication. Nate agreed to the injection and he was placed prone. The crease where his proximal hamstrings meet, his buttock was prepped and draped after isolating the point of maximal tenderness. Using sterile technique a 20 gauge spinal needle was used to inject 1% Lidocaine as the needle was inserted to the bony insertion. Then 1cc of dexamethasone and 1cc of Kenalog were injected. The needle was removed. There were no complications and he tolerated the procedure well. He did feel slightly better after the injection. We will continue to follow Nate on a daily basis.

The shot gets me back on the field but I'm a shadow of my former self. It's hard to watch myself on film. I'm slower than everyone else. My "mild hamstring strain" won't heal. Ice here, heat there, stretch here, rub there, inject here, pills there: nothing is helping. NFL athletes are so fast and explosive, that if you're not at your best, you are vulnerable on the field. It's a dangerous place to be a gimp. Desperate to get healthy, I spend more and more time at an off-site chiropractor's office. Rod introduced me to Dr. Nelson Vetanze a few years prior, and I've been going to him for periodic adjustments. He's worked with the Broncos in the past but had a falling-out with Greek over treatment philosophies. Now Nelson sees some of us in his private practice. Greek knows we see him but as long as we don't talk about it, it's not a problem. Nelson has a more holistic approach than the assembly-line philosophy used in the NFL. When I tell Nelson what they are telling me at work, he can't believe it. He knows it's not a "mild hamstring strain" but there's nothing any of us can do about it, Greek included. The only thing that would help isn't an option: rest.

But I learn to deal with the pain, the instability, the imbalance; just like every other NFL player does. My story is not unique. Every other football-playing man deals with the same cycle of injury and rehab, separated by periods of relative health. Some bodies are better suited for the demands of the game than others. They stay healthy longer, play more, smash skulls more, die younger. I should see my inability to stay healthy as a blessing in the long run, because it's sparing my brain the extra punishment. The fact is, no one will remember any NFL game I'll ever play in but me.

On our first team meeting of the regular season, Coach says that before he gets started, someone has something to tell us. Out of the corner of my eye I see a figure descending the stairs on the right side of the room. I turn to see Jerry walking down the steps wearing a tan suit. He gets down to the front of the room and thanks Coach and turns to face us: nearly one hundred grown football men sit rapt. I know what's happening. Jerry's made the team after a solid camp, but he can see how the offense is stacking up. Jerry Rice is nobody's backup. I sit in my chair and watch my hero tell me he's retiring from the game of football. No more catches, no more records, no more Jerry. I hear the voice of Bay Area commentator Joe Starkey echoing through my head:

—Touchdown 49ers!

At the beginning of every practice, before we stretch, our defensive captain Al Wilson calls someone to the front of the group to break us down and send us into the day's work. We all clap in unison until the designated breaker downer drops into a low stance and does a little dance. It's meant to be silly and slightly humiliating for whomever Al calls to the front. For this reason Al usually picks a rookie or a team clown to come up and do it. And out of respect for Jerry, Al had never called him up.

But here's Jerry, in front of us all for the first and last time. After he says all he wants to say about the end of his career, he's about to walk off when he thinks of something and stops.

—Oh, and one more thing. Since you guys never had me break us down out there, I want to do it now, in here, one last time.

He starts clapping and we follow: the whole room clapping in unison, echoing through the meeting room. Jerry breaks us down, giving us a quick preview of the moves he'll later use on *Dancing with the Stars*. The room erupts. Jerry smiles and takes a mental picture, then walks up the stairs and out of the meeting room forever. My hero has left the building. The door clicks shut and it's back to the business of professional football.

The NFL is a job. Every morning I go to work through the facility's cafeteria door, scarf a plate of whatever looks good, fill up a cup of hot coffee, and go through the locker room toward our main meeting room. Sometimes there's a note hanging in my locker signaling that I have been selected for a drug screening, a random steroid test.

Once I see the note on my locker, I have four hours to produce a sample. If I can't produce in four hours, it's considered a failed test. We have a Pee Man who sits in a folding chair near the urinals and handles all the urine. When it's ready to flow, I knock on the door to his office. It echoes off the porcelain gods.

—Hey, Nate. You ready?

—Yes, sir.

—Okay, wash your hands; water, no soap.

I wash them.

—Now select a cup.

I grab one of the prewrapped, sterilized urine bowls, unwrap it, and hand it to him.

—Okay, now remove your shirt and pull your pants below your knees.

I disrobe and he hands the cup back to me. I take my place at the urinal and he takes his place three feet away, perpendicular to me, and watches me pee. Ever since a player was caught at the airport with a prosthetic dick, called the Whizzinator, long ago, the league has mandated a visual inspection by the Pee Man, who confirms that my urine is exiting my actual penis through my very own, personal urethra.

Some players get gun-shy and have to drink copious amounts of fluid to fill the cup, only to be told that their urine is too diluted. The rules are strict. They want to catch people cheating. But I've never seen steroids around. I've never seen anyone get caught, either. I've never heard anyone talk about taking them, or heard from anyone that someone else was juicing. I've never seen HGH, either. I only learn about it through the media. They tell me it's undetectable, that it's a wonder drug, that its benefits are manifold, and that it can accelerate the healing process after an injury. Sounds great, if you want the truth, but even if I knew how to get them, I don't have the time or the peace of mind to arrange something so secret. Besides, I made it here without them. We all did. Steroids won't make you a good football player.

• • •

After my piss test I go into the team meeting room and take my seat at 8:14 a.m. for our Thursday special teams meeting. There are large, digital clocks throughout the facility so that everyone is on the same schedule. Punctuality is important; the meetings start on the dot. The clock strikes 8:15. Ronnie clears his throat, takes two gulps of coffee, and starts coaching.

—Gentlemen, listen up, okay? Hey, good work on punt team yesterday but we need to tighten up that left side when they twist. Look, gentlemen, this is our most important operation here, okay? We gotta be able to protect our punter. Or he's gonna be shitting his pants back there. Let's not leave the game up to the athleticism of our punter, okay? No offense, Micah. We gotta communicate out there, gentlemen. Okay? Let's take a look at the film.

Forty-five minutes later the rest of the team shows up. Starters don't come to special teams meetings. They're happy not to play during the ritual sacrifice of kickoffs and punts, but maybe they're also a bit envious. We're a tight-knit group. We know things the other guys don't. We know about fifty-yard dead-sprint head-on collisions. We know about snot bubbles. We look at the game differently.

And none of us played much special teams in college. In those days, we were the starters who trickled in after special teams meetings and sat down as the room's laughter trailed off from a joke we'd never understand.

At 9 a.m. Coach Shanahan comes in to address the whole team. The entrances are at the back, like a lecture hall. To get down to the front stage, you descend the stairs on either side of the room. Coach takes his place at the podium and gives us our marching orders. Here's what we have to do, let's go get it done. The head coach is the big-picture guy. He runs the show. His underlings tighten the screws. The defense has a stable of coaches: a defensive coordinator, and a coach or two for every position. On offense, Kube runs the passing game. Rico, our O-line coach, runs the running game.

The small technical stuff is left to our position coaches. That's why some of them get a little carried away with the technique stuff: it's all they control.

After Coach's spiel the defense leaves the room and Rico takes the stage and installs the run plays. About now in our daily program is when I usually zone out. We're learning the same plays . . . again. The game plan varies from week to week, but only about 10 percent of it is new. The other 90 percent is made up of the same plays we've been running for years. Even the new plays are based on the old plays. Once you learn the concepts and the terminology, everything else falls into place. But every week Rico teaches us the alphabet as if we never learned it the first time. B, it turns out, still comes after A.

After the run install, Rico and the O-line leave and we install the pass plays. Same old plays with a few new wrinkles. I have to know the wrinkles. After that I can let my mind drift again. I'm prone to flights of doodling and prose, I admit. If a coach glances at me in the middle of his dissertation, he'll think me a football scholar, quietly taking notes.

When that's over we break for fifteen minutes and Brew goes over the script for the day. Every day has a log-sheet that contains the plays we'll run that day, in the order we'll run them. It also has the defensive front, coverage, down and distance, hash-mark, and blitzes, which helps practice move along smoothly. No alarms and no surprises, please. It's mostly for the coaches. Helps them feel like they're in control of the product. Inevitably, that all falls to shit on game day. That's when you need the players to take over. To this day no coach has successfully scripted a football game.

After a forty-five-minute walk-through—really a jog-through of the plays we'll run later in practice—it's back inside for lunch. I'm never hungry. My stomach is in a perpetual knot. Every day after walk-throughs, Mike Leach whizzes by me with a nice soup and sandwich combination on his way to the players' lounge to do some emailing and eat his lunch, humming a little tune as he goes. I'm envious. If anyone has job security, Mike does. As long as he snaps well, he eats well. I eat horribly. I'm losing the weight I put on in the off-season.

And every Friday morning we have a weigh-in. They want me at least above 230 pounds. But I'm a little under. I walk into the locker room and past Crime. He sits next to the scale near the entrance, weighing guys as they walk in.

—Nate! You gotta weigh *in*.

—Okay, Crime, give me a second.

—Why do you need a second, Nate? Just do it now!

—Just a second, Crime! Jeez.

—You always do this. I don't know why you always *do* this!

I walk to my locker, change into my team sweats, sneak into the training room, and grab an ankle weight. I strap it on under my sweats and walk back to the scale where Crime is sitting with his clipboard.

—Told you I'd be right back. I just had to change.

I step on the scale.

—Two thirty-four.

—Nice. Thanks, Crime.

—Whatever.

The extra weight is hard to keep on. So even when I'm not hungry, I have to eat. I force down a ham sandwich and some goulash in the cafeteria and go into the locker room to get ready for practice.

One minute before 1 p.m. I jog on the field with the rest of the proletariat, hard hat in hand, ready for thirty minutes of special teams practice. My feet hit the grass and I flip the switch. Time to be a man. Practice moves from one scripted period to the next, with a bipolar air horn blow to signal each new period. Even though the Wednesday, Thursday, and Friday practice schedules are nearly identical, Thursday's the hardest practice day of the week because we are in full pads. We're lucky here in Denver. Coach Shanahan doesn't believe in having us beat the crap out of each other. He tries to keep our bodies fresh. Lots of teams hit two or three times a week but we only have to bang today. And nothing says banging like nine-on-seven.

Nine-on-seven is a drill for running plays. There are no receivers or defensive backs in the drill. Other than that it's a live running play. The receivers and the defensive backs are on a different field with one of the quarterbacks doing one-on-ones. When the sad horn blows to signal the start of nine-on-seven, I longingly watch my old buddies jog to the other side of the field where, for the next twenty minutes, they'll play schoolyard football. It's mano a mano over there, a receiver and a cornerback. You get the ball no matter what.

No such luck in nine-on-seven. This is tactical mayhem. In an actual game, the defense has to honor the threat of a pass. In nine-on-seven, there is no threat of a pass, so the defense explodes off the ball downhill, trying only to stuff the run. Players grunt, coaches yell, and pads and helmets crack, creating a frightening symphony of future early-onset dementia. But we have to do it if we want to be a good running team, which we always are.

Nine-on-seven is a special kind of hell for me because no matter who I have to block, I'm outmatched. I'm no tight end; I only play one on TV. I have to employ some different tactics to get the job done. One of those tactics involves a bit of trickery. Since I'm quicker than the meat sacks I have to block, I can afford to take a jab step and a head fake to get them off balance before I pounce on them. But I can only do it once in a while or it won't work. And coaches don't like finesse in the run game. Rico always glares at me when I try something different. There is only one right answer!

So I play along, shooting myself out of my stance headfirst and trying to crack the end in his face before he can move. I aim for the tonsils with the crown of my forehead. Frontal lobotomy. That's my only chance. My technique is horrible—I'm standing straight up, crossing my feet over each other, ducking my head—and I lack the size and strength to overpower anyone on the line of scrimmage. My only chance is to hit them first, pound my feet into the ground and hold on until the whistle blows. If it feels like a firecracker exploding in my helmet I know I'm on the right track. Then I go back in the huddle and do it again.

If we mess up the scripted plays, we run them again until we get them right. The period doesn't end until Coach nods to Flip, and Flip lets the horn blow. If we are having a bad day, it seems to go on forever. Crack! Internal brain bleed. Smack!

—C'mon, *Nate*!

Strange sensation in the jaw.

—What the *fuck*, Wesley!

Dizzy. Achilles tendon pain.

—Blue twenty-eight, Blue twenty-eight, Set *hut*!

Hemorrhaging contusion. Stinger. Pop. Dislocated finger.

—Huddle up!

Shit.

I keep a quick eye on Flip and his healthy finger, resting delicately on the orange button of the horn clipped to his waistband. Press it, Flip! Let it blow, Flip! *Please!* One hundred dollars. *Five hundred dollars, Flip!*

CRACK! SMACK! WHAMMO!

Eventually we get it right and Coach nods. Flip lets it blow. The same evil sound that started the period thirty minutes ago is now the sweet, sweet sound of survival. Another week is passing. Time slows down to honor its most painful landmarks, stamped in wounds across my battered frame. One more horrible thing is over.

The next morning we watch film of nine-on-seven with the whole offense. Watching it in front of my friends is torture. My shittiness is on full display. Rico leads the meeting. I remember every play from the previous day clearly, and I know when a bad one is approaching on film. My heart speeds up as the tape rolls on. Here comes the play where I get body-slammed. I hold my breath. Ooooh, that was ugly.

—C'mon, *Nate*. That's not good enough.

Stick the knife in, Rico.

—If you can't make this block, we've gotta get someone who can. Twist it, Rico.

—We can't win with that kind of effort.

Pull it out and wipe it clean, Rico. Now stick it in the next guy.

I'm stuck on the bench all season. There are fifty-three players on each team but only forty-five suit up for the game. On game day I walk into the locker room and Chris Trulove, one of our head scouts, finds me and gives me a thumbs-down sign and tells me that I'm "down." That means put on your sweats, not your uniform. We don't need you today. You suck at football.

After sulking for a few minutes, I go out to the field to run the 16x100's required of inactive players by Rich. Then I set in on the doughnuts and coffee. Knowing I'm not playing immediately relaxes my stomach. I stand and eat sunflower seeds on the sideline while my friends play. At halftime I solicit one of the ball boys to steal me a hot dog from the equipment room. I make sure to reward him with an important sideline chat. There's this thing called a "vagina." Come here. I take him over to the dry-erase board and uncap a pen. He walks away.

Bobby was right, I'm a worthless piece of shit. Compounding my feelings of inadequacy is the success we are having as a team. Jake's our undisputed leader and has us on a roll. We lost our opening game in Miami, then we won five in a row. We lost against the Giants in New York then won another four in a row. We have a complete team: offense, defense, and special teams. The locker room is a happy place. Charlie's playing a lot and catching passes. Kyle's our starting fullback. Rod's healthy and playing well. Everyone is smiling and high-fiving and having a grand old time. The city of Denver is ecstatic. I stand on the sideline and clap my hands like a fucking fanboy.

• • •

As our success increases, so does the media presence at our facility. Every day at lunch, they mill around the locker room, creating an obstacle course of people holding cameras and microphones, trying hard not to look at our genitalia. During one of the lunch media sessions, Frank Schwab, a beat reporter, asks Charlie for an interview. Charlie knows all his plays, never drops the balls, is always on time, runs good routes, and so on. He's a reliable player and everyone on the team knows it. The previous game, though, Charlie dropped a tough catch in the fourth quarter. Frank decides to turn it into a story.

—How bad did it feel when you dropped that pass in the fourth quarter?

—Did you think you let your team down?

—What did Coach say to you when you came to the sideline?

—Do you expect those opportunities to come again after that drop?

—Do you think Jake lost confidence in you after that play?

Charlie answers them all calmly and professionally. But my locker is right across from Charlie's. I'm listening to the whole thing. After the last question about Jake losing confidence in him, I interject.

—What kind of question is that?

—What?

—I said what kind of question is that?

—What, you didn't like that question?

—No, I didn't.

—I'm trying to do my job, Nate.

—Oh really? That's your *job*, Frank? Please.

—Yeah, that's what . . . I'm . . .

He stands up and stomps off. Then he pops his head around the corner.

—Mind your own business, Nate.

With that, he's off again. We laugh loudly, our cackles reverberating through the locker room and chasing him out the door.

Rod doesn't take any shit from the media. He runs the show. He only answers questions on Thursdays for ten minutes before practice. Everyone knows it so they don't bother him otherwise. At the allotted time, twenty people crowd around him and shoot questions at him, a hectic scene. Rod is in total control. If he doesn't like the question he simply won't acknowledge it, or he'll give the reporter a look that sends him or her retreating to the other side of the locker room. I have a front-row seat for the Rod Show because his locker is next to mine. I have to climb over cameramen and reporters to get to my stuff. To make it extra awkward for them, I wait until they arrive before I change. Nudity is the one power move that the media have no answer for. Sure, you can ask that question when my clothes are on, but how about . . . now?

Our offensive line doesn't speak to the media at all. They have an internal policy enforced by fine and degradation that no offensive lineman is to speak to the media at any time for any reason. It's common knowledge around the locker room. The media members know it and never bother asking any O-linemen any questions, unless there's a new guy or a rookie. Then they try to catch him before he learns the rule, just to get him in trouble.

As the season progresses, our playoff prospects come into view. If we win our division we'll have a home playoff game no matter what. If we have the best record in the AFC we'll have home field advantage for the whole playoffs. We travel to Dallas to play the Cowboys on Thanksgiving. It's a big game and we're very focused. But we also have penises.

My friend has a lady there that he wants to see. But the hotel for away games is like Fort Knox. No nonplayers can get on our floor. And we can't leave the floor after 11:15 p.m. So he books a separate room at the hotel that will serve as their lion's den. He puts it under her name and she checks in and waits for him. He scopes out the exits and decides that a stairwell on the east side of the building will be his best bet. There's security at every exit so it amounts to choosing the friendliest-looking guard. He tells the rent-a-cop that his fiancée is in the hotel and she's pregnant and he wishes he could spend the night down there with her. Then after bed checks he goes back to the same exit and asks the security guard if he'll let him pass.

—No dice.

—Really?

—Nope. Sorry.

—*Seriously?* She's pregnant, man.

—Sorry, strict orders.

—One hundred dollars.

—No.

—Two hundred dollars?

—Nope. Can't do it.

—Wow.

He goes back to his room and calls her and explains why a grown man is not allowed to leave his hotel room. He tells her he'll get down there as soon as he can. He goes to sleep and sets his alarm for 5 a.m. When it goes off he puts on his clothes and goes to the elevator for "breakfast," nodding to the prison guard.

—Good *morning!* Can't wait to eat *breakfast!* Really been thinking about those pancakes all night. Gosh, I'm hungry.

Ding.

He gets off at the lion's den floor and slides into bed with her for as long as he can before he has to catch the late bus. He tells me the story at the stadium and I'm tickled. All of this talent, all of this work we go through to get here, all of the sweat and the blood, and a man can't even get laid without cutting through miles of red tape. We beat the Cowboys. Ron Dayne breaks a 55-yard run on the second play of overtime, setting up the game-winning field goal. I watch the game in sweats. The sunflower seeds are delicious.

• • •

The next week we go to Kansas City to play the Chiefs. I walk into the locker room and get the old thumbs-down once again. It's very cold outside. My sweats go back on and I pace the sidelines trying to stay warm. I stand in front of the industrial heater and drink hot chocolate. Then one of my inactive friends hands me a Gatorade bottle with white athletic tape wrapped around it.

—Go on. This'll warm you up.

I take a swig. Cognac. I pass it back. He laughs at the face I make. By the fourth quarter he's drunk, running all the way on the field between plays, stomping around and cursing at opposing players. No one notices. Probably because there's blood splattered all over the sidelines.

Running backs coach Bobby Turner got a bad nosebleed before the game started. Our team doctors did what they could but it only got worse. They urged him to relax in the locker room but he wasn't listening. Bobby T's not going to sit in the locker room while his boys are out there going to war. We need him. He's a calming force on the sidelines. Most coaches freak out on game day because they can't control anything. But Bobby T is calm. He is also in charge of substitutions on the sidelines. All of the offensive skill position players gather around him and await his word. Before every play, he yells out the personnel group—Base, Tiger, Zebra, Eagle, Trey, Empty—which determines how many receivers and running backs and tight ends are on the field for the play. So he stands on the sideline in the Kansas City cold with a cup full of blood in his hand and does his job.

In a last-ditch effort to stop the bleeding, the doctors stuff gauze up his nose. There's dark blood splattered on the sideline. He's coughing it up because it's backed up through his sinuses and it's drowning him. I stand next to him and keep watch. Then he blinks heavily a few times and blood trickles from the corners of his eyes and runs down his cheeks. I nudge our physician, who's standing next to me. He shrugs. Yeah, he sees it. But what's he going to do? Bobby T's a grown man who knows what's important to him. And the game rolls along, oblivious to the pools of blood.

It means a lot for me to see Bobby T bleed. As a player, it sometimes feels that the coaches who radiate toughness are only play-acting. They don't really mean it; they never have to prove it. But on the sideline in Kansas City, Bobby T proves it with every bloody tear.

We lose 31–27. It's our third and last loss of the season. After that we win four straight, finishing the season in San Diego against the Chargers, our division rivals. John Lynch sacks Drew Brees near his own end zone late in the game, dislocating Brees's shoulder and sending the Chargers into the off-season, Brees to the New Orleans Saints, and us to the postseason with a 13-3 record. As we pull out of the stadium, the Chargers fans in Black-Fly sunglasses and brand-new powder-blue jerseys, drunk and stoned on hydroponic weed and Percocets, line our path and urge us to go to hell. Really they're just checking themselves out in the reflection of the bus.

We're the second seed in the AFC, behind the Colts. After our first-round bye, the Patriots come to town, the two-time defending Super Bowl champions. They have considerable moxie. They have chutzpah. They have Bill Belichik! And Tom Brady! We have Champ Bailey. He picks off Tom Brady in the end zone in the fourth quarter and returns it 100 yards for a tugboat, sending the Patriots home shy of their quest to three-peat and us on to the AFC Championship game against the Steelers, who have just beaten the Colts in Indy. As Champ runs alone down the sideline, [Insert Corporate Logo Here] Field explodes. It's the loudest I have ever heard the Denver fans. We're one win away from the Super Bowl.

Almost immediately my phone starts ringing. Everyone I know wants to fly out for the game. Eight friends end up coming, even though I assured them I won't be playing. That's not the point, I soon realize. They want to come to Denver and party.

I'm living in a one-bedroom apartment with an office. Four of them stay at my place and four get a hotel room around the corner.

There are empty bottles and pizza boxes and odious dude smells clinging to the ceiling fan after just one night in town. I'm thankful to have some peace and quiet in the hotel the night before the game.

The morning of kickoff, I come home and pick them up and we drive downtown. I laugh as they recount the previous evening and make a mental note to steam-clean my futon. The drive into the stadium is a psychedelic trip on an orange-and-blue carnival ride. The Broncos are back and Denver is in a state of euphoria.

But in the locker room there is a different vibe. Since I'm not playing, I'm hyperaware of how the locker room atmosphere translates onto the field. I usually know if we'll win before the game starts. I sit in my locker and savor my chocolate doughnut. I know it will be my last of the season. The locker room is dead.

We come out flat and go down 24–3 in the first half. It's all we can do to claw back and make it interesting, but we lose 34–17. It was never even close. My mind flashes to my eight friends who insisted on coming even though I wasn't playing. In addition to them I got another fifteen tickets for Denver people. I can only imagine the friends and family of my teammates who are actually playing the game. Charlie's family from Pennsylvania was staying in his house with him all week. It was the same way for everyone. We were stretched thin by the love. By the time the game came around we were wiped out. Sometimes home field advantage can be a distraction.

The Steelers, who had squeaked into the playoffs, go on to win the Super Bowl. The airwaves light up in the days that follow our loss. Despite three straight post-season appearances, the fans are unhappy. The big finger points at Jake Plummer. Yeah, I guess it was a good season. But not good enough. Now someone has to pay.

6

Plummer's Crack
(2006)

The next week Gary Kubiak is hired as the head coach of the Houston Texans. It was only a matter of time before he got a head-coaching gig. Kube was a rising star in the coaching world. But Jake and Kube are very close. Kube has served as the buffer between Coach Shanahan's offensive philosophy and Jake's unwillingness to be stuffed in a box. With Kube gone, something has to give. As the draft approaches, the talk of the town is not about our team's success; it is about Jake. Coach trades the 15th and 68th picks to St. Louis for their 11th pick in the draft. Roger Goodell walks to the podium and delivers the news.

—With the eleventh pick in the 2006 NFL Draft, the Denver Broncos select Jay Cutler, QB, Vanderbilt.

The local media rejoice. They are still under the hypnotic effect of a powerful sedative, street name: JEN. John Elway Nostalgia isn't fading with the years. It's growing. There is Joe Montana nostalgia in the Bay Area but it's different. Joe was traded when his health began to deteriorate. Steve Young was given the job and didn't miss a beat. They're both in the Hall of Fame now. They both won Super Bowls for the city, with contrasting styles of play. Niner fans learned several valuable lessons about the NFL through the trade of Joe Montana. One, there is no loyalty. Two, there are no storybook endings. Three, there is more than one way to play quarterback.

John Elway was the exception to almost every NFL rule. He was one of the most highly touted college athletes ever. He was the first pick in the draft. Then he played sixteen years for the same team. He avoided major injuries. He went to a million Pro Bowls. He won league and Super Bowl MVP, respectively, the latter in one of his five World Championship games. He won two rings, at the ages of thirty-seven and thirty-eight. Then he retired. That won't happen again.

But in Denver, it fucking better!

The drafting of Jay is met with a powerless shrug in the locker room. We have no quarterback controversy. Jake is our guy. But we could see it coming. And it only serves to hammer home some points that most of us already understand. The NFL is not your friend. Your job is not guaranteed. Your life is under scrutiny. You play until they replace you.

We all have our own position battles to worry about, too. Jeb was released in the off-season, freeing up a tight end roster spot. I got excited for a moment, then the draft happened and I regained my perspective. After Jay, in the second round, we draft tight end Tony Scheffler. Then Brandon Marshall, a wide receiver, then defensive end Elvis Dumervil, wide receiver Domenik Hixon, and guard Chris Kuper. In addition to the draft, we trade for standout wide receiver Javon Walker. For every man who comes, one must go. Someone is always getting the old squeeze.

Tony is a pure pass-receiving tight end: freaky size and speed and great hands. A quiet, dark-haired, six-foot-five Michigan native, he is carrying the hopes of his family and his town and his school on his back. Add Brew to that load. Wes has injured his knee in our first minicamp, freeing up all of Brew's time for Tony. Brew is mainly a run-blocking coach. He has only a nominal interest in the passing game. His bailiwick is biting someone in the fucking neck! He likes to get dirty. By that I mean that he likes for us to get dirty. Tony was a wide receiver coming out of high school and got bigger in college so he switched to tight end. But he has a receiver's skill set. That's what attracted scouts to him and that's why we drafted him in the second round. But Brew wants to turn him into a road grader, a no-nonsense smashmouth run-blocking monster truck. "Attitude and effort are nonnegotiable," Brew always says. Once we get to training camp and put on the pads, he shows Tony what he means.

—Tony, do you like how you look on film?

Silence.

—Well, do you?

—No.

—Fuck, you better not! You look like a fucking pussy, Tony. Look at this shit. You call that a fucking block? Stick your fucking face in there. You look like you're playing patty-cake, Tony. Look at the weeble-wobble off the line of scrimmage.

—What the fuck is that? Be a fucking man, Tony!

—Fuckin' bite him, Tony! Stick a fuckin' knife in his hip! Tie his dick in a fucking knot, Tony!

—Oh, fuck me gently, Tony. What the fuck is this?

S.A., who played for Brew in San Diego, reminds Tony after meetings that Brew doesn't even hear the words that he's saying. He doesn't realize he's being a dick. In his mind he's being tough but fair. But I can see it in Tony's eyes: he's taking it all personally. He's chained to the wall in the dungeon of Brew's sadistic football mind. Every day Tony's eyes are getting wider and sinking farther back in his head. He is losing weight. He is cracking up. He needs to find a ladder and get the hell out of the tank.

Every night at our seven o'clock team meeting, we start things off by having a few rookies entertain us. One at a time they're called to the front of the room to be harassed and humiliated. They must state their names, position, school, salary, anything anyone yells out, etc. Then they entertain us with a joke, a song, an impression, any-thing. If the performance sucks, he is booed off the stage and must return another night until he gets it right. A good performance wins your freedom, and more important, it wins over your teammates.

Tony, Jake, S.A., and I are eating dinner before meetings one night. Jake asks Tony if he has a joke ready. He says he has a few ideas but isn't sure. He runs them by us. The one we like the most is the one he's most reluctant to tell. We tell him that's exactly why he has to tell it. But he's not sure. He thinks he's going to tell the one about the priest and the push-pop. Jake tells him that would be a mistake. Go with the one about the genie in the bottle.

In fifteen minutes we are all in our seats as the digital clock hits 7:00. We start the drumroll on the desks. Coach Shanahan yells:

—Tony!

He looks at me and stands up, then squeezes behind me, S.A., and Brew on his way into the aisle that takes him down the stairs and up to the podium. The team fires questions at him.

—Who are you?

—Tony Scheffler, tight end, Western Michigan.

—No! Who are you?

—Oh . . . Rookie.

—What was your signing bonus?

—Umm . . .

Boo and hiss.

—Okay, I have a joke for you guys.

He glances at Jake.

—So I'm out on the field the other day at practice and I missed a block, and I look over at Brew and he has that look on his face. You know, that *look*.

Chuckles from the audience.

—And when I walk back behind the huddle, he grabs my face mask and says 'Tony! What the fucking fuck are you fucking doing, you bugfucker?! Don't you know what a fucking *power block* is?'

Audible laughs.

—So needless to say, I'm feeling like shit after practice, walking through the locker room, and I kick a water bottle, and a genie pops out. The genie tells me that I have one wish. One wish? Man, I'm thinking to myself, I really just want to get drunk and forget about everything. So I say to the genie, 'I wish that I can piss vodka.' The genie shrugs and says, 'Okay, and so it shall be,' and disappears in a puff of smoke. I go to the bathroom and take a piss and sure enough, pure vodka! So I go into our meeting room and I tell all the guys about it and everyone gets excited so I fill up everyone's glass and we're having a grand old time. Well, most everyone forgets about it the next day, except for Brew. He comes up to me after practice and says, 'Hey, boy! How 'bout some more of that vodka?' So I say okay and I fill up his cup again. And the next night, Brew finds me again and grabs me by the back of the neck. 'Hey, boy! How 'bout some of that good stuff?' So I hook him up again. Then the next night, I see him walking toward me in the hall. He had just got through moth-erfucking me out on the field again. And here he comes with that big ol' smile on his face, and he says to me, 'Hey, there, boy! How 'bout some of that good stuff?' like I'm his best friend in the world. So I said, 'All right, Brew, I'll give you some of that good stuff. But tonight . . . you're drinking from the bottle.'

Yes! The room collapses in laughter and applause and apprecia-tion of the rare moment when a player regains the upper hand. Brew laughs along and takes it all in stride. He may be a hard-ass, but he has a sense of humor. Tony walks back up the aisle and squeezes be-hind Brew, S.A., and me, and sits back down as I congratulate him.

—Well done, bugfucker!

Coach Shanahan descends the stairs with a smile on his face.

—Good one, Tony. We'll see how that one plays out tomorrow at practice.

Late the next week I see Charlie in the hall before meetings and he tells me he's just been traded to the Dallas Cowboys. He says he's leaving, like now. His flight is in a few hours. I don't know what to say to him. Football goodbyes are strange. It's like he's being deported, voted off the island, banished. It happens nearly every day to somebody. And I know from experience, it's likely that I'll never see him again. NFL players evaporate when released.

He is on the plane to Oxnard, California, that night, which is where the Cowboys have training camp. He's met at the airport by a driver who takes him to the team's headquarters. The next morning he takes his physical, which is a formality before he can get on the field. But he does not pass the physical because of his knees. He's had multiple ACL tears and knee surgeries over the years. His knees are junk. He practices every day in Denver on those knees, but they're not good enough for Dallas, and the trade is nullified. He gets back on a plane to Denver and is back in meetings the next evening as if nothing ever happened. But now he knows the score. With Rod, Javon, Ashley, and now Brandon Marshall, there is a surplus at wide receiver. Someone is always getting the old squeeze.

Camp winds down and preseason games are in full swing. Jay has overtaken Bradlee for the number-two spot. Bradlee is doing his best to keep calm and carry on but the Jay Cutler tidal wave is rolling over everyone in its path. Jay plays well in the preseason, which disproportionately excites the media and the fans. But Jake plays well, too.

He throws me a 35-yard touchdown in the third preseason game against the Tennessee Titans. I'm wide open after a nice play-action fake by the entire offense, and as the ball floats down into my hands, I have to concentrate extra hard so as not to drop it. The wide-open catches are the hardest because you're thinking about catching the ball. Catching the ball is an instinct: a reflex. When you stop to think about it, you put a kink in the circuit. Empty your mind and you'll catch everything.

After the game, Charlie, Kyle, Matt Mauck, and our friend Grant Mattos stand at midfield and talk. Matt is gone now. He was cut the previous season and now plays for the Titans. Grant's from San Jose, too. And we both have Ryan as our agent. Grant went to the University of Southern California, then San Diego, then we signed him in the off-season. He merged into our group of friends immediately but he was cut the day before training camp started and landed in Tennessee with Matt. Now here we all are, reunited for a moment on the fifty-yard line. We pose for a picture and say goodbye, scattered again to chase our gridiron dreams alone. The next week our roster is set.

Despite the media's Jay fetish, Jake is our starter for week one. It's in St. Louis and we put in a very heavy game plan. Nearly every team, every year, freaks out this way because the coaches have had all summer to prepare for the first opponent. None of the preseason games mean a thing. Game one looms as soon as the schedule is set. Game one matters. When it finally arrives coaches want to fire all of their guns at once.

Going into the game we have multiple audibles and line-of-scrimmage checks that depend on what defensive fronts and coverages and blitzes Jake sees when he gets under center. If they show this we change the play to that. If they show that we change it to this. If they blitz this guy we'll do this, that guy we'll do that. But it will be loud down on the field. So we practice in front of high-powered speakers blasting white noise so we can get used to our silent snap count. At home you can listen to the quarterback's cadence and move when he says, "Set-hut!" But when you can't hear him you have to go when the ball moves. If you can't see the ball you move when everyone else moves. This slows down your jump by a count because when the quarterback says, "Set-*hut*," you really fire your gun on the soft and rolling "Set" even though the hard "hut" is emphasized. "Set*hut*" is said as one word, and since the offense knows the snap count and the defense doesn't, the offense starts moving before the defense and catches them off guard.

Crowd noise takes away that advantage. It also makes it hard to change the play at the line of scrimmage. We can't hear Jake's audibles so we have to look for hand signals. A tight end in a three-point stance can't always see hand signals, so we have to look at the defense and know how Jake will change the play based on what they're showing. Football players are smart and all, but it's not our main thing.

As expected, it is very loud in St. Louis on game day. The audibles and the blitz-reads are too much for us to handle. We never get anything going. We lose the game 18–10 and have five turnovers on offense. Jake throws three interceptions. In the locker room after the game, some reporter (probably Frank Schwab) asks him if he thinks people will be clamoring for Jay now.

—I'm sure they will. They've been calling for him since he got drafted.

Jake's candor is rare and doesn't help his image in Denver. But he doesn't care anymore. By then the two—Denver and Jake—have fallen out of love with each other.

One bright note from the game: Rod went over 800 career receptions; the only undrafted player to ever catch that many passes. He is thirty-six years old and he's the savviest receiver I've ever watched. He understands the angles better than anyone. And he also understands the simple concept that many ballcarriers often forget: the end zone is north and south, not east and west. He catches a pass and shoots like a rocket straight up the field, always tacking on at least five yards to a catch that normally would be stopped on the spot. And he'll play until his body breaks.

Jake is right about the clamoring for Jay. The following week the airwaves light up with JEN-induced Jay love. But after the loss in St. Louis, we go on a run. In the middle of that run, we travel to New England to play the Patriots. On Saturday morning, right before we leave for the airport, Charlie gets a tap on the shoulder again. But this time he isn't being traded. He's being cut. And no strange goodbyes either. I don't notice his absence until the plane takes off and Charlie isn't in his seat. Later, dude.

We end up winning five in a row and are alone in first place, and dating back to the previous season, we are 19-5. But still there is dissatisfaction with the product; with the way we are playing. Simply winning is not enough.

We lose to the undefeated Colts at home 34–31, and the "bench Jake" chorus starts up again.

—Look at Peyton Manning! Now that's a quarterback!

He threw for 345 yards and three touchdowns. Jake threw for 174 yards and one touchdown. These stats weigh heavy on the minds of Broncos fans and serve as a smoke screen to the view of our team's success.

After our loss to the Colts we go to Pittsburgh with revenge on our minds from the AFC Championship loss, and we get it. Ben Roethlisberger throws the ball 54 times for 433 yards, one touchdown, and three picks. Jake tosses it 27 times for 227 yards, three touchdowns, no picks, and no credit. Javon Walker and our defense are celebrated for the victory.

The next week we beat the Raiders in Oakland. I'm starting to figure out the tight end position. S.A. is our starter and Tony is our main passing threat but I'm getting some good action. In the second half I catch three passes in front of my friends and family, all dressed incognito so as not to rile up the indigenous creatures at the Oakland zoo. Four of my close friends have some nice seats down in the north end zone. They are dressed in neutral colors, as directed, and are staying silent all game. When I catch my first two passes, they keep quiet. They have made friends with the surrounding Raider fans and have bonded over an assumed mutual hatred of the opponent.

But in the third quarter I catch a pass on a corner route that leaves me running down the sidelines toward my friends. I am tackled around the five-yard line and they instinctively jump up and high-five each other. Whoops. They've revealed themselves as the enemy. And not just the enemy: the enemy in disguise! They spend the rest of the game deflecting thrown trash and idle threats. They learn a valuable lesson that day in Oakland: it's a good thing the bottles are plastic.

• • •

As the season rolls along, I use my free time to go house hunting. I've been in Denver for over three years, and live in a nice apartment in a cool eight-story brick building ten minutes from the facility. But lots of guys are buying houses in the suburbs. Me, too, I think. Why not?

Things have started to fall apart with Alina. She didn't know anyone when she first came to Denver a few years earlier but by now she has a group of party friends who are pulling her in all directions at once. I want no part of them, especially during the season. She wants fun and excitement. I have a playbook to look over and my body hurts. You want me to go to dinner and talk with these people? About what? The Broncos? I'm staying home.

Then when I tell her I'm thinking about buying a house she sits me down and tells me how to do it. She insists that I use her mother, who is a real estate agent in California, as a reference to my Colorado real estate agent so that she will get the referral fee and kick it back to us. I don't want to do that. I like my real estate agent. I don't want any funny business. Alina can't understand why I'm being so hardheaded about this. She tells me I have no idea what I'm talking about. This is how it works. To spite her, I go about it alone and close on a house in Greenwood Village, a suburb twenty minutes from downtown and four minutes from Broncos headquarters.

A few weeks later we break up. We could both see it coming for too long now. Three and a half years of young NFL love is over. We never stood a chance. I move into my new suburban family home alone: 2,600 square feet of future regret, with a fabulous view of Cherry Creek State Park and the best neighbors a guy could ever have.

Back at work, the Denver media are orchestrating their coup. They've gotten what they wanted—a loss at home to our division rivals, the Chargers—and now the drumbeat comes louder and faster, drowning out our 7-3 record.

Local shill Mike Klis of the *Denver Post*, on November 22, 2006, the day before our Thanksgiving game against the Chiefs:

"Just in time for Thanksgiving, it's open season on Jake Plummer. The whole town, it seems, is in an outrage."

The next day, on Thanksgiving morning, national shill Adam Schefter (Denver's former local shill) hit the wire with "breaking news":

" . . . Jay Cutler will be starting for the Broncos on December third against Seattle . . . a Broncos team source [says] that Cutler would've been starting this week if it wasn't such a short week for the Broncos."

We lose the game in Kansas City. Obviously. The gallows lever had already been pulled. It's hard to play quarterback with a noose around your neck. After the game we slouch in front of our lockers removing our gear and tape. We had no mojo. Our usually potent rushing attack was stifled all day long, gaining only thirty-eight yards, and our usually stout run defense gave up 223 yards. Jake played pretty well, I thought. It was the team that lost the game.

The media are giddy as they enter the locker room and make a beeline for Jay's locker, which is next to Jake's. The backup quarterback doesn't get interviewed after games. But they want to crown Jay right there in the Kansas City locker room with the grass stains still on Jake's ass. And they want Jake to see it. That's the moment when I permanently lose faith in sports media. They don't give a fuck about us. They want to watch us burn.

After they're done with Jay, they set in on Jake.

—Have you heard anything from the head coach, Jake, regarding your . . .

—No. I haven't heard anything. I get little bits and pieces from people around me, ya know, when people are saying, 'Hey, hang in there. Don't listen to what's going on.' I realize that it's the media, really, you guys, that start that stuff because it's your job to, and, the best I can, I shut it out because I know I have a lot of fans that are rooting hard for me. Yeah, there are some who don't want me to play anymore, but I can't control their thoughts unless I play well.

—What's the most frustrating part of the situation?

—Not winning ball games. That's it. I don't care if I play like shit. I want to win. That's all I care about. I don't care how pretty I look, obviously.

He points to his shabby outfit.

—I want to play ball and try to win games for my team, and if that doesn't happen, that's frustrating to me. A lot of times I get too much credit and I get too much blame. Right now, the blame is there. I didn't make some plays today. And I've got to make those plays.

—Do you think you will be the starter next week?

—Did I just not answer that question for you? I don't know. I'm taking three days off. You guys will probably know before me because I don't read anything, I don't listen to all you guys on the radio, I don't watch any of your TV shows. When I find out, I'll find out. Whatever it is. And if I'm starting, I'll bust my ass as hard as I can, for Al Wilson, for Rod, for all those guys. That's how I play, that's what I've always done.

—Do you think you deserve to be the starter?

This last question, another gem by Frank, really gets me thinking. What comes first, Frank, the chicken or the egg? The story or the storyteller? Did you create the need for the story or did the need for the story create you? And why can't you just be cool for once?

Either way, a few days later Jay is named our starting quarterback.

Privately a few players grumble, but for the most part everyone stays silent. We know we don't have a say in it. We felt the pressure weighing down on the building from the day Jay was drafted. It was the elephant in the room and it took a dump everywhere. We were stepping in elephant shit on our way out to practice every day. The media had JEN in their eyes every day; their every question was laced with it. The only thing that can equalize JEN is a story of what could be, not what is, because nostalgia is, in its way, an unwillingness to accept the present. That's why they love Jay so much, because they haven't had a chance yet to decide that he's not John.

An NFL football team is not built to depend on one man. It is built to rely on one system. The men are temporary. The plan is permanent. The scouting department brings in the talent, and once they're in that front door, they become cogs in a machine. Jake has never been benched in his life. Confronting the reality of the machine is something he hasn't had to do until now. Franchise quarterbacks are the last bastion of sentimental aw-shucks football fairy tales. Former quarterbacks and quarterback coaches wear suits on television and tell football fans why the quarterback is all that really matters. But someday that quarterback will be thrown out with the trash. Eventually the lie reveals itself to everyone. Everyone except John.

Seattle comes to town for Jay's first start. It's to be his grand entrance against an inferior opponent: a perfect first game for a rookie quarterback.

But we lose 23–20. Jay's play is understandably erratic. Jake's demeanor is understandably aloof. Coach Shanahan has made his decision and there's no turning back, come what may. But it's the middle of the season and we're used to a certain game-day vibe and style from our quarterback. It takes time for an offense to adjust to a new one.

First, the two of them throw very different footballs. Every quarterback gives a personality to the ball he throws. Each one is a snowflake. Wobble, spin, angle, trajectory, velocity, accuracy, timing: all unique to the thrower. This information is vital to the receiver. Know thy ball and ye shall catch thy ball. Some balls are misleading and tricky, come in at strange angles, fall like torpedoes, wobble and break. Some balls are pearls. Some are rainbows that shoot from the quarterback's hand. The receiver's hands are the pot of gold.

Jay's ball came nose down with an aggressive spin. Jake's was nose up and a little softer. Unless you catch it clean with your fingertips, the ball's movement will determine its ricochet, which in turn determines how a receiver positions his body for a ball that's coming in hot. Knowing where the ball will come down before the defender knows where it will come down is 90 percent of the battle as a receiver. If I react first to the ball in flight, meaning, if I understand the ball's flight better than you, then I will be there sooner, and will create a wall between you and the ball with my body. Now all I have to do is catch it. Nose-down ball means it is diving and I need to get my hands underneath it. Nose-up means it's rising and I need to get my hands on top of it.

But we understood the difference in their balls from practice. The main adjustments were game-day stylistics. How does he feel? What does he like to do? What does he see when he scrambles? What parts of the field does he like to exploit? What do his looks mean? What routes does he prefer? How does he communicate? This stuff comes along slowly.

The next week we go to San Diego and lose to an especially game Chargers team. We are powerless to stop LaDainian Tomlinson's mojo. He scores three touchdowns in the game, which gives him twenty-nine for the season: an NFL record. The crowd chants "M.V.P! M.V.P.!" He'll go on to play eleven seasons in the NFL, quite a feat for a superstar running back. Running backs have very short careers. The better they are, the more they're used, the faster they fall apart. The human body can't absorb that punishment for very long. A thirty-year-old running back is a rare sight in the NFL. LaDainian will play into his thirties and walk away before someone tells him he has to.

We are on a four-game losing streak now: two with Jake, two with Jay. We need a win to stay in the playoff hunt and we get one in Arizona. Jay settles into his role. We need another win the following week, at home against the Bengals on Christmas Eve, to even think about the playoffs. The media are optimistic about our chances. They have the QB they want. They excuse Jay's mistakes as growing pains, comparing his first few games favorably to those of John. The JEN drips off the page.

But then a blizzard hits. On Thursday morning, I wake up and can't get out of my house. There's a four-foot snowdrift in front of my garage door. I'm alone in my suburban family home. Everyone else will be late to work, too. Eventually Tony comes and picks me up in his Hummer. He can't make it into my neighborhood so I trudge the quarter mile to his car, with the snow coming up to my knees. The air is calm. The snowflakes fall in slow motion. The world is silent except for the crunch of my boots. Peace is everyone stuck indoors.

We make it to the facility and sit in empty meeting rooms and watch film. Most of the team doesn't get there until lunch.

The coaches are pissed! So pissed! Fuckin' snowy, fuckin' icy, cold frozen watery substance in the clouds! Fuckin' clouds! Who are the Clouds, anyway? Don't they know we play the Bengals in a few days?

Whenever the big weather hits, we bus up to the street to an indoor field we call "the bubble." It is only eighty yards long and is not as wide as a regulation field. I like going to the bubble because it breaks up the routine. After practice we bus back to the facility and fall back into our normal meeting schedule. Jake has accepted his fate and seems more relaxed than I've seen him in years.

Despite our lack of adequate preparation we win our next game 24–23. Carson Palmer drives the distance of the field in the final minute to score a touchdown that gets the Bengals within one point. The extra point will tie the game. But the same storm that crawled over us a few days earlier whipped us with its tail on its way out of town. The long snapper loses his grip, and the holder can't get it down in time. The elements are relevant. We win.

● ● ●

That puts our record at 9-6 and with a win or a tie we'll make it into the playoffs and save a tumultuous season. Our opponent is the 49ers. They're 6-9. It's a cold New Year's Eve day in Denver. There is snow on the ground and a chill in the air. I say hello to a few old friends during warm-ups. Things have gone south for the organization since I left. But they tell me they've got something special for us. They're going to send our asses home. We'll see, motherfuckers! We'll see.

On a punt return in the first half I initiate contact with my man on the line of scrimmage and try to sustain my block down the field but he's too fast and I have to sprint to keep up with him. He's running straight for our returner Darrent Williams, with a step on me already. Just as D-Will catches the ball I give one last burst and try to position myself so I can block my man without clipping him. I dip my shoulder and push him hard, a borderline block in the back. He misses Darrent but flies into Curome Cox, one of our defensive backs. Gets him in the legs. Curome crumples to the ground and grabs his knee. I'm off balance and run straight into Darrent. He bounces off me and runs up the sideline for a 34-yard return.

Curome lies on the ground, attended to by Greek, Boublik, and Corey. Eventually he gets up and walks to the sideline. He's okay, after all. I tell him I'm sorry. Then I have a laugh with Darrent.

—I thought you were tryin' to tackle me, Nate! Damn!

—I was!

—Ha! Already!

D-Will always says "Already." It can mean anything, depending on the situation. Usually it means that everything is well and good. You already know.

Jay takes a shot to the head in the ensuing drive. He's woozy and has to come out of the game. Jake goes in for a series and throws an interception, much to the chagrin of the JEN crowd. Still, we go into halftime up 13–3. Jay shakes off the vapors and reenters the game in the second half. After a touchdown pass by Alex Smith, Jay throws an interception of his own that is returned for a touchdown. The life is sucked from the stadium with an audible hissing sound, like a balloon deflating. Without a specific quarterback to blame, the desperation becomes a heavy blanket of inevitability that suffocates us on the field.

The game goes into overtime. Twice we have the ball and twice we have to punt from our own territory. That means twice I have to block their special teams ace coming off the edge trying to stuff the kick. I am the wing on the right side: one yard off the line of scrimmage and one yard off the ass of our end. I knew all week long that I would have my hands full with this dude. On film he was a beast. He blocked several kicks during the season and always made the wing look stupid. He has a three-step up-and-under move that is very good but that I know I can stop. But knowing and doing are much different.

We line up for our first punt in overtime. He digs his feet into the dirt and gets his ass high in a sprinter's stance. I picture the worst-case scenario: He runs right through me, blocks the kick and returns it for a touchdown. Game over. Have a nice life. The thought makes me sick. I dig my feet into the dirt, lowering myself in my stance. I can see his breath coming out from under his helmet. The ball snaps and he fires out: one-two-three steps and cuts hard across my face. I'm there to meet him with a good pop.

—Ahhh!

He laughs and mumbles something as he peels around the backside and chases the ball down the field. Fifteen minutes later and we have to punt again. The punt is identical to the last one; except this time Alex Smith drives the Niners' offense close enough to the end zone and Joe Nedney hooks in a field goal to send us home for the off-season. The balloon flutters flat to the frozen grass.

I wake up at eight the next morning to a phone full of missed calls and texts. I am groggy from the night out after the game. I call Kyle.

—Kyle, what's up, man? What's going on?

—You don't know yet?

—Know what?

Pause.

—D-Will got shot last night.

—What?

—He's dead, Nate.

He's dead.

The words rip through my head, tearing apart everything I know to be true. From the moment he was drafted two season earlier, Darrent was a light in the darkness. He was a spark of energy in the locker room. He had a smile for everyone. Fuck football. He was a good person.

As the day unfolds, I find out what happened. After the game, while some of us were at a bar called Spill, another large group of teammates and friends and family were a few miles away at a club called the Shelter. There was a fight in the street after the club let out and security broke it up. Some of our group jumped into a Town Car, but most of them got into a white Hummer limo. The guys they were fighting ran off and got into a white Ford Bronco.

A few blocks from the club, the white Bronco pulled alongside the white Hummer limo. The driver reached across his passenger and opened fire on the Hummer with a .40-caliber handgun. The limo was full and the music was loud so no one heard the shots.

D-Will was sitting next to Javon Walker, who was keeping track of D-Will's chain through the scuffle. Javon was in the middle of a sentence when D-Will slumped over into his lap. Javon was confused, laughing midsentence and trying to pull D-Will back up. Then the windows shattered and everyone heard the shots and dropped to the floor. The limo pulled off the road. The Bronco sped off. Two other people were hit, neither of them badly.

But D-Will was hit in the jugular. Javon pulled him outside of the limo and tried to resuscitate him but the wound was too much. He bled to death in seconds. Darrent Williams died on New Year's morning of 2007 in the Denver snow, ten hours after playing in an NFL football game.

• • •

We all instinctively show up at the facility when we find out. We sit in our lockers and look around the room bewildered. Someone tell us what to do. Tell us this isn't really happening. I look at D-Will's locker: the locker of a working football player. His jerseys and sweats and sweatshirts are on hangers. His helmet is on the hook. His shoulder pads are on the shelf. Pairs of shoes, lotion, tape, papers. I know he'll come walking around the corner at any moment and sit down at his locker with that smile on his face. He isn't dead. None of this happened. Nothing can get into our football bubble. Nothing gets past this locker room. D-Will grew up in a rough neighborhood in Fort Worth, Texas, and while some of his buddies got into the street life, his football talent kept him out of it. He's a father, a son. His life is just starting.

We are supposed to be at the facility for our exit physicals anyway, but the protocol crumbles under the weight of the moment. We are deaf, dumb, and blind. In our football lives, we pretend we are invincible, because we have to keep on playing. In reality, we are fragile and we are afraid.

The next day we have an informal memorial service at our facility. People share their stories, their thoughts, and cry with one another. Darrent's mother, Rosalind, flies in from Texas and comes to grieve with us. She amazes me. She is so strong, so composed in the face of tragedy. We are falling apart all around her and she stands up in front of us and tells us, a group of one hundred grown men, frightened men, that everything is okay. Everything is going to be okay.

Our last team trip of the year is to Darrent's funeral in Dallas–Fort Worth. Our entire team and staff make the journey. The service is all power and glory and spirit and raw emotion in a packed evangelical church. They welcome us with love. I sit motionless on the wooden pew looking at the pictures of D-Will in the program for the service. I knew him. But I didn't know him like these people knew him. I wish I had.

After the service we file back onto the bus. Jake is sitting in front of me. We look out the window as D-Will's loved ones walk slowly out of the church, some crying, some smiling, some at peace, some not. Jake turns back to me.

—Man, Nate. You know what I can't stop thinking about? I mean, if I would have played better on Sunday, you know? When Jay got hurt for those two series and I went in, I mean, if we would have won the game.

—What do you mean?

—You know what I mean. If we would have won the game then we would have made the playoffs and guys wouldn't have been out partying like that, and . . . I don't know, I just can't stop thinking about—

—Man, you gotta stop that. There's nothing anyone could have done.

—I don't know, man. I don't know.

We sit in silence. Football is nowhere to be found. Life swallows us whole. The bus pulls away from the church.

7

Pointy Balls
(2007)

What is it with the pointy balls?

—It makes them easy to throw. They spin through the air and travel really far that way.

—Uh huh. And what about all of that armor you wear? Is that really necessary? I mean, isn't that a little, I don't know, cowardly? Are you boys afraid of a few bumps and bruises?

—We get plenty of bumps and bruises. The armor makes it more violent. It's a dangerous sport *because* of the armor. The helmets are made of hard plastic and metal and that allows you to use your head like a spear. You can hurl yourself at your opponent with no regard for your body.

—Well. I'm not sure I like *that*. Okay, one more question. What is it with all of the starting and stopping and the discussions and conferences you all seem to be having with each other? What could possibly be so important?

This one is more difficult to answer. I'm sitting in a loft just off Las Ramblas in Barcelona, Spain, with a few new friends: Andy, Dan, and Lucy. A Scotsman and two Brits. After D-Will died I sank into a hole. When I crawled out I came to Europe. The NFL off-season is about three months long if you don't make the playoffs. It goes by very fast. I want to try to enjoy this one. I'm single and free for the first time in my career. But I'm confused about my life as a pro athlete and embarrassed that I feel this way. You're living the dream, boy! What do you possibly have to complain about?

We pass around a spliff and trade culturalisms. They think it's something of a novelty to be chatting with an American footballer in Barcelona. Especially Lucy. She has plenty of questions. The dream isn't so dreamy to her, either, the more I explain the situation. My friend murdered, my body in pain, my nerves fried, my relationship exploded, my life a series of yes-sirs. And a few days earlier another teammate had died unexpectedly. One of our running backs, Damien Nash, collapsed shortly after playing in a charity basketball game and died of heart failure. He was a happy, optimistic player working his way into the league. Another vibrant young athlete is dead. The palace walls are closing in.

The NFL bubble is well formed. It keeps almost everything out: everything but the big stuff. When tragedy intrudes no one knows what to do. We are ill-prepared for life. We don't know how to handle our emotions.

—Yes, but Nathan, darling, that isn't an American football problem.

—You don't think?

—Certainly not. It's the same thing in my life. Everyone is always so distracted, simply cannot be bothered. And when something happens, or someone is honest for once, or someone shows a moment of vulnerability, they're punished for it.

—Yes! Because people see it as a weakness that they're trying to convince themselves they don't have, so they suppress it themselves by rejecting it in others. But where does it go? It goes somewhere, right?

Andy has a look on his face. Then he speaks with his thick Scottish accent.

—Into our dungeons . . . where it *rots*.

Lucy agrees with him.

—Yes, into the dungeons, with the rest of our real emotions. Our emotions are the only things that we truly have, that are truly ours, and we are taught to reject them.

—Damn. You're right. So what do we choose instead?

—Darling. Pointy balls, of course.

I come back to America determined to excavate my dungeons. The first thing I learn is that Jake has retired. He is thirty-two years old and healthy. Still a star quarterback, he's chosen principle over promise and left the industry that betrayed him. The second thing I learn is that I have a new tight end coach. Brew has accepted the job as the head coach of the University of Minnesota Golden Gophers. Taking his place is Pat McPherson, Jake's old QB coach. Pat is the son of legendary 49ers coach and executive, Bill McPherson. After a solid college career at Santa Clara University, Pat jumped into the coaching cauldron. He is working his way up the ladder, a pace made slower by the fact that he's not an asshole.

Soon our off-season conditioning program starts up again, minus number 16 and the "bugfucker." It's the first week of April. The off-season program is "optional." That means if you decide not to do it you'll be fired. But under the bylaws of the collective bargaining agreement, teams can't technically require players to participate. They can just strongly hint at its importance. There also can't be any coaches. It's just the players, plus Rich and Crime.

There are five workouts a week. If we make it to four out of five, it's considered 100 percent. We're paid $110 a day to work out. And they serve breakfast and lunch. Compared to actual football practice and meetings, it's fun. I like it. All I have to do is show up and work hard.

It's a two-part workout: the run and the lift. The run is the hardest. It's usually broken up into three or four stations, which vary from day to day: straight-line sprinting, agilities, quick-feet, hurdles, shuttles, ladders, box jumps, ball drills, and an ingeniously concocted torture contraption: a stable of wooden sleds.

Rich oversees the whole operation. He loves his job and he loves us. Well, most of us. That means that there's no amount of torture he would require of us that he wouldn't require of himself twice over. The wooden sleds are the best example. They look similar to hurdles but are much larger and much heavier and have a platform on them upon which Rich can stand and yell in the face of the sled pusher if he wants. But he doesn't do that. They're heavy enough already.

We have to push them about forty yards across the field. This is done at the end of a workout involving a few other stations. The cherry on top is a set of twelve. One rep is painful. Twelve is hell-fire. The lactic acid builds up so violently in the legs that the skin feels ready to explode. Guys are laid out on the ground panting, vomiting, and crying. Each teardrop that hits the ground fertilizes Rich's torture garden. He loves to break us down.

Rich regularly pushes twenty or thirty sleds in the mornings before we arrive. On his fiftieth birthday, he came in early as usual. But instead of his normal workout, he honored the landmark by pushing fifty sleds in fifty minutes. That would kill an ox. Rich didn't break a sweat. I am always excited to hear what crazy-man workout he did while I was still sleeping. But soon I notice a pattern developing. The harder his workouts get, the harder ours get. If a fifty-year-old man can do it, then surely eighty professional athletes in their twenties can, too.

Only that isn't always the case. Football players are dynamic athletes. We burst. We explode. We do not plod. Our slow-twitch muscles are not refined and when pushed, they crumble. Rich enjoys this immensely. We bitch about it every day. But we need the slow-twitch work. He promises us that if we win the Super Bowl, we can have a bonfire and torch the fuckers. Sorry to spoil the ending, but we never do win that Super Bowl.

Show me a painful ritual and I'll show you a way to cheat it. The mischievous O-line discovered that the sleds slide better on wet turf, so whether we're indoors, on the forty-yard synthetic grass field connected to our hangar-size weight room, or outdoors, enterprising slackers squirt down their tracks with water when Rich has his back turned. This makes life much easier. But look out for Crime! He's the enforcer.

—Don't water down the grass, Nate!

—C'mon, Nate! Touch the line!

—All the way through, Nate!

—C'mon, Nate! That's only nine!

—How many times are you going to tie your shoes, Nate?!

Crime is a loyal and humble assistant. Together with Rich they make a daunting duo on the field and in the weight room, where we each have our own folder with our daily lifting regimen inside. Weight lifting is important. But everyone's body is different. Every body responds differently to the strain. Some guys get too big, too bulky, too weight room strong, and have no mobility on the field. When I moved to tight end I started lifting much heavier. I gained twenty-five pounds. I was much stronger. I lifted incredibly hard. And my body fell to shit.

Champ is a naturally freaky athlete. He does not have to lift weights. One off-season he participated fully with the lifting program, grunting and pressing and pleasing Crime tremendously. All parties agreed: Champ looked better than ever. But by the time the season came around, his body was giving him trouble. He battled injuries all season long, his least healthy as a pro, and learned a valuable lesson that only a guy like Champ could actually pull off: weight lifting isn't really my thing, coach.

But most of us don't have that luxury. We lift and run as we are told. No one asks us how we feel. It's assumed that we feel fine and that we are ready to push on, harder and faster. There simply isn't time to pay attention to the individual athlete's body. It's the industrial football complex. Here's the program. Go.

But it isn't bad. I'm nitpicking. My job is to work out. I like working out. And it only makes me feel better about going to Las Vegas, a trip that is already in the works. I've been going to Vegas since college, when I had little money and few resources. Just another random dude. If you haven't been, the random dude experience in Vegas goes like this:

Four guys jump in a Toyota Corolla and drive seven hours from San Jose to Las Vegas and check into their room at Bally's. Two double beds, two dudes to a bed. They throw their stuff in the room and go down to the casino immediately, hop on the five-dollar blackjack table, and look around for the cocktail waitress. Where the fuck is she? Five dollars a hand and free Coronas until forty to sixty dollars is lost.

Twenty minutes later they're walking down the Strip with plastic yard-margarita souvenir cups and three-and-a-half-foot straws. Men on the streets hand out cards advertising prostitutes.

—*Vegas showgirls at your door in fifteen minutes!*

—Holy shit, bro! I'm calling this girl later. Look at her!

He slides the card into his pocket, to be fingered the rest of the day as they wander the Strip ogling the scenery and playing low-stakes blackjack. One of them is winning. The other three are in the hole. Two have already gone over budget for the trip.

They are all drunk. They stop at McDonald's in the Paris food court for dinner. They throw french fries at each other. They meet a group of girls near the bar of the casino. The girls are there for a bachelorette party. They are all sucking on dick-shaped lollipops and they have dick-straws in their margaritas. The bride-to-be looks frightened. She is wearing a crown and a sash and has a list in her hand. She decides that one of the dudes is a good candidate for the "Let a random guy take a shot out of your belly button" item on the list. He chooses tequila. The Cuervo stirs up fermented navel bacteria, forms a sexy cesspool. She laughs and it spills down her sides. He slurps it up and fist-pumps. This will be the highlight of the trip.

They say goodbye to the girls and go back to the room to get ready for the night. Tequila-shot guy passes out. The other three shower and change into their best shirts. Tequila-shot guy won't wake up. They leave him and decide to go to a popular club. They wait in the cab line for thirty minutes and arrive to a hectic scene.

There are several long lines to get into the club. Beautiful women are everywhere. The guy with the showgirls card thinks he sees the girl on the card walk past him. He waves. She does not.

They get in the general admission line and wait. They watch people being ushered in past velvet ropes: packs of girls, groups of dudes, shaking hands with bouncers, laughing. The general admission line isn't moving. They try to get a bouncer's attention. He pays them no mind. They look at each other. Fuck this. They go to the casino bar and order drinks. They lose one friend to the blackjack table on their way to the bar. He will not be seen the rest of the night. The two remaining random dudes meet another bachelorette party. The rest of the night is a blur: more shots of tequila, low-grade narcotics, lots of walking, a few cab lines, and an after-hours club full of dragon chasers. The two remaining random dudes stumble out of the club into the morning daylight and look around. They cannot find a cab. They don't have money for one anyway. And they have already reached their withdrawal limit for the day.

They walk back to the room and fall on the beds with their friends, taking care to keep a layer of blanket between them. They sleep until 2 p.m., wake up, and repeat the previous day's drill. On Sunday they get back in the Corolla and sit in Vegas exodus traffic for the next four hours. Tequila guy pukes out the window, and gets some in the car. They all tell him to go fuck himself. He swears them off as friends, puts on his headphones and pulls on his hood, fakes sleeping. He thinks about his tequila shot.

—Did anyone get a picture of me and that girl?

No one answers him. They ride home in the silence of their self-loathing.

What our Vegas experience as NFL athletes lacks in Cuervo, it makes up for with Patrón. First I go to the bank in Denver and get a large wad of cash. I learn that from Rod. I want to know what I'm spending. The teller smiles knowingly.

—Going to Vegas again?

—Yes, ma'am! But this time I'm bringing some back. In fact I'm gonna double it.

—Yeah, right. Have fun!

The NFL has scattered our friends around the country. This Vegas trip is a chance to get back together. Patrick Chukwurah, a muscly, dreadlocked defensive lineman who we hung out with in Denver, is in Tampa now. Grant is back home in California. Charlie is in Houston with Kube. Kyle and I are still, for the time being, Broncos.

We touch down in Vegas and retrieve our bags. The cab line is too long, we decide. We walk to the other side of the terminal and find an idle SUV. One hundred dollars to the Wynn. Free bottle of water in the cup holder. Definitely worth it.

We check in at the VIP desk. We all have our own rooms and plan to meet downstairs in thirty minutes. It is wise to ease yourself into your Vegas weekend. Don't walk in the front door of the casino and go straight for the roulette table. Check in, put your bags down, and relax. Have a beer. Go to the pool. Look at the boobies. There will be plenty of time to hate yourself later.

We are very happy to be in Vegas. We splash around in the water and make friends.

—What do you guys do?

—You play for the Broncos?

—*What?!* Honestly, like, the Broncos are *literally* like, my favorite teeeeeeam!

—Oh, my *Gawd*! Let's take a picture!

Our cocktail waitress is pretty and easy to talk to. I like the way she keeps her bottle opener tucked into the side of her red bikini bottoms. As the sun arcs above us, we become best friends. She likes me for me, I tell myself, not for the excessive bottles and shots I'm ordering.

We meet a bachelorette party in the pool. The bachelorette considers Charlie for the "random football player she'll cheat on her fiancé with" item on her list.

The sun is almost down. The pool is almost closed. We're the last group. The cocktail waitress sits down next to me as we close out our bill.

—What are you guys doing tonight?

—I think we're going to Tryst. Do you want to come?

—Maybe!

—Maybe?

—Yes, maybe. Give me your number and I'll text you later if I can come.

We all go up to our respective rooms for some quiet time before the evening. The rooms at the Wynn are nice and spacious. I throw my board shorts on the couch and walk to the floor-to-ceiling window. Las Vegas: so beautiful, so ugly. I plan the evening in my head, lie back on the bed, and doze off.

A knock at the door from room service wakes me up. Pepperoni pizza, chicken fingers, fruit, side salad, water, six beers, service charge, delivery charge, casino charge, resort fees, utensil rental, tip: $112.67. Whatever. It's just my signature.

We meet downstairs and play blackjack. We look for the cocktail waitress. Where is that lovely woman? I can't wait to meet her. Oh there she is.

—Hello, Edna. A Tanqueray and tonic, please.

I win nine hundred dollars in thirty minutes. Give forty to the dealer, forty to Edna. The cocktail waitress from the pool texts me.

—Hey! Me and my girlfriend are coming. You boys better be nice!

Just before midnight we go to Tryst. We spot Ryan, our VIP host. We all shake hands, then he pulls up the velvet rope and we walk in, past legions of random dudes waiting. I lock eyes with myself, seven years younger. I look restless.

—We have you guys at a great table. You're going to like it.

I put down my credit card. Ryan ushers us past a few more ropes and lines and we walk down the staircase and into the long hallway of the club, where the lighting changes from paltry to sultry.

Ryan was right. It is an excellent table: near the dance floor but not too close. Charlie slips Ryan a few large bills. He leaves us, and I turn to my brethren.

—Fellas, what do we want to drink? It's a three-bottle minimum.

—Three bottles? Dawg, are we gonna drink all that?

—We have no choice!

—Vodka.

—Vodka.

—Vodka.

—Tequila?

—Yeah, tequila is good, too. We'll start with those two and worry about our third bottle later.

Grey Goose and Patrón: bottle-service booze brilliantly marketed to the tune of $475 apiece. It's the price of having your own table and couch: your own private island in a sea of sleaze.

Our waitress introduces herself. She's a typical Vegas industry girl: hypersexual, overproduced, worn-out. I give her our order.

—What mixers do you guys want? OJ, tonic, soda, Red Bull, cranberry?

—All of the above.

—Okay. Do you want some water?

—Yeah, six.

While she's gone the bouncer in our area introduces himself.

—Fellas, you look like you don't need any help but if you do, let me know. Anything you need, I got you.

Handshakes and hugs all around. These people really love us.

Our waitress from the pool arrives with her friend. She looks much different done up and dressed to go out. No bottle opener and no bikini but she looks very good. I pour them vodka and Red Bulls. We yell into each other's ears from inches away. I ask questions. She gives answers. We chuckle with tight lips.

After the arbitrary get-to-know-you conversation, I push through the haze of smoke and bad decisions and go to the bathroom. When

I return, women have emerged from the fog, pulled toward us by our oversized pituitaries and our caveman libidos, vibrating the floorboards like a Dr. Dre bass line. The music pulses through the high-octane speaker system and into my bones. I lean back on the cushy couch and watch. Who are all of these girls? I don't think they actually exist. The paper in our pockets has conjured them out of thin air. And now everything is open wide: arms, doors, and legs. We are young, physically powerful men with money. Big money usually doesn't come quickly. When it does, it's rarely because of physical prowess. We are temporarily rich *because* we are bigger and stronger than you. This unnerves people constantly. Well, it unnerves men. It nerves women.

A few tables away a group of Englishmen start throwing money in the air. Dead presidents are pinwheeling around us in the current of the high-powered air conditioners. The Brits are jumping up and down on their couches yelling. They are spraying champagne.

A table of Persians is not to be outdone. They start throwing money in the air, too: but they use larger bills. The whole club stops to honor the moment. One of our new friends walks over and scoops up a handful of cash and comes back to us. She drops the money on our table next to several nearly empty bottles of booze.

—That's for later.

Girl from the Pool and I look at each other and laugh. Her red lips and white teeth shine in the blue-black backdrop. I want to kiss her. Our hypersexual waitress has started drinking with us. Now she's dancing with us. Now she's giving Kyle her number.

—You better call me! What are you guys doing tomorrow? There's this cool bar off the Strip that we should go check out if we have a chance.

It is 3:30 a.m. We have been through five bottles. Our bill is over three thousand dollars. I tip the waitress five hundred. Fifteen percent really shouldn't apply to bills like this but who cares. I will collect the money from the boys later if I remember. We skip into the casino with ten new friends.

—What's the plan? Is anyone tired?

A sweet harmony of "No!"

Cab line is too long again. We find a limo and we get in. A hundred dollars for a half-mile trip to Drai's at Bill's Gamblin' Hall. More free water. Turn up the music. Sing songs of the damned. Pull up to the curb. Tip our driver an extra twenty. He gives me his card. There's a small mob in front of the entrance to Drai's. We walk to the front of the line. Ryan is a host there, too. He opens the rope.

—You guys want a table? It's just a one-bottle minimum. Otherwise it's pretty crowded down there.

—Yeah, let's do it.

Another credit card changes hands, another sacred gate unlocked. We descend the stairs to the vampire dungeon. Googly-eyed bobblehead dolls float across the sticky dance floor, pause to light cigarettes, can't steady the flame, close one eye, finally get it lit. The bathroom attendant broods over a glass vase full of bills. He squirts soap into my hand. Dries my hands off. Zips up my fly. I drop a five-spot into the vase. We bump fists: best friends forever.

I walk around and engage in one-liner small talk with approachable vampires. I feel the wad of cash in my pocket. It wants to be spent and I want to spend it. I want to feel the bills peeling off one at a time, slowly lightening the stack. I want to pass them along and keep the Vegas food chain strong. I want to help the economy. I want to spend my money to remind myself that I have it, to remind myself that I am special, that I am desirable, that I am somebody.

The rest of the night dives into the gutter.

We fly back to Denver when we've had enough. The taste on my lips makes the self-loathing easier to swallow. The paper in my pocket is gone.

After a few hard days of working out, the Vegas fog is lifted and only the beautiful memories remain. I have been texting Girl from the Pool since I got back to Denver. One day, after having spoken the previous day, she texts me and tells me she is having phone problems and will be using a different phone until she gets it fixed. Later that day, she amps up the flirting and asks if I'll send her a picture. "You first," I say. She sends me a couple of innocuous pictures of her: one at a table with a "Happy New Year" hat on, one posing outside near a bush. So I send her a few innocuous ones of my own.

The next day I wake from a nap to a very long voice-mail message from her. She says that she is married. Well, she's separated. It's a long story. But her husband has gone through her phone and discovered our flirty texting. He then texted *me*, pretending to be her, and gave me the broken phone story. The number that he redirected me to was his own. I sent pictures of myself to her husband, after he sent pictures of his wife to me. Oh, she was sorry, so sorry about this; so, so sorry. I soak my phone in bleach and get back to work lifting heavy pieces of metal.

We are done working out by noon every day. It's a very good feeling showering after a hard workout, sitting down for a free lunch and looking up at the clock, knowing I have the rest of the day to do whatever I want. It's the ideal stoner schedule, really. Wake up early, shake off the cobwebs of last night's fun with some exercise and have the rest of the day to kill. But NFL stonerdom is a more calculated endeavor. The off-season months, which might be used to make ganja-induced epiphanical deposits in the bank of the soul, instead are spent abstaining in anticipation of the league's once-a-year street drug test. By the time the draft comes around, you'd better be good and clean, because the testing starts during minicamp. Like Greek said years ago, if you can't pass this test, you're either stupid or you're an addict. Either way, you need help.

They test one position group at a time. Sometimes it comes in May, sometimes in June, sometimes in July, and sometimes not until August. The August test is a real buzzkill. That means months and months of thumb-twiddling and gazing off into the distance, enticed by nothing but raging hormones. Stoners are content sitting on the couch and thinking. Nonstoners need actual action to pacify them. They need booze and sex: or God. God is usually the odd man out. The NFL should remove marijuana from their banned substances list. Don't tell anyone about it: just stop testing for it. Pain is a big problem in the NFL. Pain management is necessary. Weed is the least harmful and least addictive of the painkillers players use to cope with the violent demands of the game. Drug use in the NFL mirrors drug use outside of the NFL. Pills reign supreme. There are more overdoses in America from prescription painkillers than from cocaine and heroin combined. And no one ever overdoses from weed. The problem is pills and booze. A joint can alleviate the need for either and plant buttocks firmly on the couch, where a *MacGyver* marathon takes on epic proportions. And no one gets hurt, except for the idiot who locked MacGyver in the bowels of a sinking ship.

The week after the draft is our first minicamp. At the urging of their position coaches, the rookies transcribe every phrase uttered during meetings, assuming they are all important. That used to be me. They'll learn soon enough. You don't need to write down a word. The constant drone is by design. It seeps into your brain regardless.

Out on the field, minicamp practices are fast and crisp. Coach wants us practicing at game speed. If you practice slow, the game feels too fast. If you practice fast, the game feels slow. Knowing *how* to practice is just as important as knowing *what* to practice. But the rookies are lost in both regards. Not only do they have to figure out the tempo and the nuanced contact/noncontact line that we toe every day, but they have to learn a new language extremely fast. Each offensive system is a foreign language. Its cornerstone terms have no meaning outside the system. The terminology is dictated quickly and with the assumption that it is understood, even when there is no way it will be. We install the entire playbook in the first few days, knowing full well that it will whistle through the ears of most everyone, because during the second minicamp, it will be installed again, and again during training camp, until it becomes second nature. Until then, there will be confused rookies getting yelled at for fucking up play after play, day after day. I'm happy to be done with it. My grasp on the offense is complete. No surprises and no confusion. I can just go out and practice hard, then come in and hope the Pee Man is ready for me, so I can go home and watch *MacGyver*.

In early June Coach Shanahan hosts his annual golf tournament. It's a popular event, drawing wealthy and influential people from all over the country. It usually falls during the week, and Coach gives us the option of going to the tournament or working out at the facility as scheduled. If we go to the tournament we get credit for the workout.

And if you go to the tournament you don't have to play golf. You can just ride around in a golf cart and schmooze. Those of us who play, maybe twenty of us, are put in foursomes with corporate bigwigs or local heroes or cookie-cutter rich dudes. There are gift bags at the check-in desk, sponsored holes, gourmet food, beer tents, endless prizes, video cameras, and an awards dinner at the end of it. Golf balls are stacked into pyramids at each tee on the driving range. The beer is free and fully stocked at every hole.

Cigars are lit and we tee off. It's a scramble, or "best ball" format, so my poor golf skills are protected. I'm the "Broncos player" in the group. Although they would have preferred one of our big-name guys, they'll take what they can get. And once their disappointment with their pairing wears off, they feel great knowing that no matter who I am, they're probably better than me at golf. Coach Shanahan always says that if a player is really talented at golf, it's not a good sign. I find this encouraging. We have a few decent golfers on the team, but they are always either quarterbacks, kickers, or snappers: finesse guys. For the most part, football players swing a golf club like they are trying to make a tackle. That doesn't translate to a good golf shot. Our foursome bonds over my inability to relax and swing the club. We toast our beers and pull on our cigars. I act like

I know how to smoke one. We get drunk and tease each other. Does your husband play? Ha!

It's a fun event, but what pushes it from "fun" to "no way I'm missing it" is the driver assigned to each player's golf cart. It's a Broncos cheerleader. This is a quality perk for the paying customers as well, as they are often pervy men who fancy themselves, in the right setting, able to properly love a beautiful young cheerleader, if only things had gone a little differently.

For me it's a chance to have an actual conversation with a dancing figment of my imagination, a princess of my football dreams, a temptress in chaps, a goddess of the gridiron. I never knew which cheerleader it would be. It didn't matter. They are all perfect.

Our final minicamp of the summer is in early July. By now the rookies have started to figure out the offense. Practices are crisp. Coach has us practicing on two separate fields simultaneously. One field is for the first- and second-team players. The other is for the rookies and the guys stuck farther down on the depth chart. Both fields run the same plays off the same script, and when we watch the film, we watch the starters' field. The rookies have to stay later or come earlier so they can watch their practice on tape. I'm on the starters' field and I'm getting lots of reps. Tony is nursing a broken bone in the outside of his foot: a common injury in football. We run a lot of two-tight-end sets. It's called our "Tiger" group. When you have fast, athletic tight ends on the field it puts the defense in a bind. If they decide to cover us with linebackers and safeties it makes them vulnerable in the passing game. If they put nickel corners on us it makes them vulnerable in the run game.

I'm working in the first group with Daniel Graham (D.G.), our new free-agent acquisition from New England. When Daniel got to town the first thing he asked me was if I would be willing to part ways with 89. He had been 89 his whole career and was hoping to keep it that way. I didn't care much for 89. It was the only one available when I switched to tight end. And since then, a few new numbers had opened up. I was happy to give up 89, but as is customary in the league, it would come at a price. I put on my hardball negotiating face and gave him my first demand: thirty thousand dollars.

He laughed in my face. "Fifteen. That's it." Fifteen is industry standard.

I laughed back. "No deal." I walked away knowing I had him right where I wanted him. When he didn't run over immediately and cave, I approached him with a new offer.

—Okay, fine. You wanna play hardball? Twenty-five.

—No, Nate. Fifteen.

Cool as can be.

—You're not moving off that number?

—Nope.

—Well neither am I. No deal.

I walked away again and we didn't speak the rest of the day. The next day I walked up to him in the weight room.

—Okay, buddy. You drive a hard bargain. Twenty. Otherwise no deal.

—Fine. No deal.

I waited ten minutes for him to realize his blunder then I approached him again.

—Seventeen-five.

—Nope.

—Sixteen-five.

—Nope.

—Sixteen.

—Nope.

—Fine. Fifteen it is.

We shook hands and the deal was done. It was strange at first watching film after practice because my eyes instinctively went to my number, 89. But by this last minicamp of the summer, my eyes have adjusted. I'm 81 now. And I like it.

8

Farewell, Bronco Betty
(2007)

The short break goes by in a blink. First I fly to Vegas and spend a few days with my new girlfriend, Sara. She lives there. A friend introduced us on one of our mancations in June. It was lust at first sight. We've been flying to see each other every chance we get. But Las Vegas takes on a different hue when you're hanging out in the suburbs, filling up gas tanks and walking the dogs. For a partying tourist, the Strip is a wonderland. But it has a runoff. Driving through desert tract-home neighborhoods with rocks for front lawns and cacti for shrubs and tumbleweeds blowing through childless streets, the city reveals itself. People voluntarily come to Vegas to live a sequin fantasy: dollar signs shooting off like fireworks in the night sky. But the sun rises and illuminates the lie. The vampires scatter like roaches into soon-to-be foreclosed tract homes and shut the blackout shades. Life is but a dream.

After Vegas I go to New York to meet Charlie and Kyle. We are on a whirlwind partying circuit, spending our money as fast as we can, chasing a ghost that is pulling away from us. But we won't go down easy.

In late July, Kyle and I go back to Denver and Charlie goes

back to Houston. Training camp starts and it's here-we-go-again. But the locker room feels different without Jake. His presence had been a part of my daily life. We had lots of talks over the years: about life, football, music, idealism, power, control, authority. That he had found success as an NFL quarterback made me believe that there was room for an iconoclast in the cloistered institution of big football. It had made me believe that a free mind could flourish in the confines of structure and rigmarole. When things fell apart between him and Mike, another piece of my idealism went with it.

But the good/bad thing about football is that it moves too quickly for your conscientious objections to keep pace. It pulls you along by sheer force. No sooner have I made peace with his absence than I'm back fighting to keep my job. I feel as healthy as I have in a while and, more important, I finally feel comfortable as a tight end. The hell I endured as fledgling blocker has given way to a polished attack that utilizes my strengths and camouflages my weaknesses. My missile-shot pop to the chin of the defensive end is refined to a laser beam. The firecracker explosion in my helmet is a familiar friend. The "Oh shit" moment is an "Oh well" moment. I'm comfortable in hell.

But less than a week into training camp, between practices, I receive word that Bill Walsh has died. I knew he was sick but I didn't know the extent of it. It was leukemia that killed him, the same disease that had taken his forty-six-year-old son five years earlier. Bill had kept the severity of the worm to himself, not wanting to worry those who loved him. His death produces a powerful reaction in the football community.

I feel instant remorse. I haven't spoken to him since I shook his hand in his office four years earlier and left for my new life in Denver. I never reached out to say thank you, to tell him what his influence had meant to me, the life that it granted me, the dream it fulfilled. I had plenty of chances to call him or write him and tell him how I felt. But for whatever reason, I didn't. Whenever the thought came to my mind, I dismissed the idea, thinking that he probably didn't want to be bothered, didn't want people's sympathy, didn't want to be reminded that he was dying.

As I walk onto the field for afternoon practice, Coach pulls me aside and asks me if I have heard the news. I say I have. We stand in a silent recognition that only Mike and I can share. The three of us form a triangle inside of which my NFL life has taken shape. It was Mike who answered Bill's call and agreed to bring me in for a look. I had been an extension of Mike's appreciation for the roots of his own football philosophy, and aside from the two of us, no one knew much about my connection to Bill. I didn't talk about it to anyone. It happened under cover of darkness. I had come to Denver alone. But Bill was always on my shoulder.

After practice the next morning, Mike asks me if I want to go to Bill's funeral service at Stanford. I hadn't thought about it. We're in the throes of camp, but I can tell that Mike wants me to go. He can't leave his football team in the middle of camp. Bill wouldn't have approved. But he wants to send a representative contingent. The day of the service, Coach charters a private plane for John Lynch and me. John played for Bill at Stanford in 1992.

Instead of waking up and going to practice, I wake up and put on a suit and drive to Centennial Airport, a mile away from our facility. John and I board the plane and sit back for the two-hour flight. We are picked up in San Francisco by a car service that drives us straight to Stanford. We check in at a desk outside the church. It's a beautiful sunny day. My suit hides my training camp welts and bruises and my blisters rub on my dress shoes as John and I find seats near the far left in an open pew. The service unfolds like a dream.

Dr. Harry Edwards, longtime 49ers consultant and iconic sports sociologist and a mountainous man with a baritone voice, takes the pulpit to eulogize his mentor and friend. Bill had known he was going to die so he prepared the service like a game plan. But he left the substance of the eulogy up to Dr. Edwards. I sit transfixed by the doctor, pulled into his words as he puts Bill's legacy in the proper light. I knew Bill Walsh as a man who helped me though he did not have to, a man who cleared the path for me to chase my dreams. I knew him as the coach of my childhood street football games. I knew him as the reason I'm a professional football player, as the reason I believed I could be one in the first place.

I scan the room as Dr. Edwards speaks and I see the look on everyone's faces. It's the same look, all of us thinking the same thoughts about the man who changed our lives. There is Joe Montana. There is Jerry Rice. There is Steve Young. There is Eddie DeBartolo, 49ers owner. There is Ronnie Lott. There is everyone—all players, coaches, politicians, family, friends—hearts opened to Harry Edwards as he contextualizes Bill's legacy. Bill Walsh was a football visionary. But he was much more. He was a social innovator in a sport that was bogged down in oppressive traditions.

He started the minority coaches internship program, which initiated the hiring of black coaches in a sport dominated by black men. And when he observed the real-world shortcomings of men who had been bred and groomed to play football, he tried to even the scales. He started the college reentry program, the postcareer occupational and preparational internship program, the financial counseling program, and the family and personal counseling program. He saw the need for an improvement in the life skills of his players and he acted on that knowledge because he loved them. He loved the athlete: not just his body, but also his mind. He wanted the athlete to flourish and achieve his true potential, in football and in life.

Dr. Edwards fights back tears and sums up the sentiment. Then one after another, a who's-who of Bay Area football greats take the podium and try to put into words what can't be: Montana, Young, DeBartolo, Senator Dianne Feinstein. The gratitude flows from pew to pew, reflects off the stained-glass windows and illuminates the photo of the Gray Fox that sits perched above a bouquet of white flowers next to the pulpit.

After the service, I stand alone outside while John talks to some old friends. I look around at all of the hardened football men I know from television, and they melt before my eyes into human beings. The 49ers of my youth stand in a cluster and embrace each other like only brothers can. This is what football can do. This is what it means. It's not the yards or the touchdowns or the money or the fame: it's this.

As I walk around the outskirts of the church I run into Doug Cosbie and Fred Guidici, two of my old Menlo coaches. Fred brought me to Menlo eight years earlier. Doug introduced me to Bill two years after that. Those Menlo years were special for all of us. I haven't seen either of them in a while. Can't remember the last time. Football is strange that way. When you're on the same team, practicing, preparing for games, sitting in a meeting room, you're as close as can be. You know everything about each other. But when you move on, you drift away. We all hug and share a few words about Bill. Then Doug asks me how camp is going, how I like being a tight end. I chuckle and he knows what I mean. He was an All-Pro tight end with the Cowboys. He knows all about it.

John and I ride in silence back to the airport, absorbed in our thoughts. A few hours later the airplane touches back down at Centennial Airport. We make the short drive back to the facility and are in our seats in time for our 7 p.m. team meeting. After the meeting I find Coach Shanahan. I thank him for the gesture. He didn't have to do that. He didn't have to do any of it. He must have learned that from Bill.

—Nate, come help me tie my tie, man. Just like yours. That knot is *clean*.

I got lucky with this knot, which has brought me Cecil Sapp's admiration. I give his own my best shot but it ends up in a silk puddle under my Adam's apple.

—Sorry, man. I can't do it.

Someone else will help him. It's Saturday morning of week one and we have a plane to catch. We are headed to Buffalo. I got back from Bill's funeral and had one of my best camps as a pro. Compared to Brew's House of Vitriol, Pat's bugfuckerless meeting room is cordial and calm, allowing us to flourish on the field. Our position group is solid. We are playing well and we know our shit. We are all veterans and we all get along. Tight end has given me a new perspective on the game, and it's made me a much smarter football player. As a receiver, I was always stuck out on an island. I learned to embrace the solitude. But we are landlocked as tight ends, in the thick of a pass-rushing, run-stuffing defense. And it's turning me into a man.

But earlier this week, the day after our last preseason game, the reality of the business showed its fangs again. Once again I played in the last preseason game against Arizona, assuming I needed to play well to keep my job. And once again I sat envious as the starters were excused from meetings and told that they weren't playing in the game the next night: they were safe. One of the safe guys was Kyle. He was our starting fullback. Had been for the last two seasons. Kyle stood on the sidelines in his sweats and his jersey with the rest of the starters and laughed while we grunted and popped our way through a meaningless game. I was proud of him. He deserved it, a security I had never known. But the next day he got a call from the facility and was cut. Even security wasn't secure. No one was safe.

We finish our morning meetings and have an hour to put on our suits and get on the buses. With the knot secure and my hair in place, I walk out of the locker room, across the parking lot, through the weight room, and into the small indoor field it's connected to. On the indoor turf stand Transportation Security Administration employees armed with wands who check us for box cutters, explosives, and liquid exceeding 3.5 ounces. After the terror test I walk out the back door to the parking lot containing five or six buses. I get on bus number 3 and open a breakfast sandwich I poached from the rookie DBs. The end of the week signals the beginning of a three-day junk food binge leading up to the game. I know there is no way I can ever overeat. I'll still be under my target weight. After everyone is aboard, the buses rev up and follow our police escort through the streets of Dove Valley and onto C-470. The motorcycle cops attack their duties with gusto, darting across lanes of traffic, yelling and pointing at confused drivers, zigzagging with menacing throttles and speeding up ahead to clear all traffic to the shoulder. Move over! There's a *football* team coming through.

Thirty minutes later we pull onto the tarmac at Denver International Airport, right next to the airplane. The wind blows through my heavily gelled hair, rustling the lapels on my oversized suit as I ascend the stairs of the airplane.

We always have the same rotation of flight attendants. They are the queen bees in the United flight attendant hive. And on this flight they don't have to go through any of the standard preflight safety instruction mumbo-jumbo. Everyone knows God loves the NFL too much to crash one of its planes. They also don't have to enforce the FAA's draconian passenger guidelines: seat back up, seat belt on, electronics off, bags under seats, no congregating in the galleys, no yelling obscenities or throwing grapes or looking at nudie magazines.

I say hello to the ladies on the way in and find my seat, labeled with a sticker with my name on it. The plane is big and spacious. Coaches up front in the luxury seats. Operational staff, media, marketing, equipment guys, trainers, etc., are crammed in next to each other in the middle cabin. We are in the back. I stretch out and listen to some music and pretend to read a book.

After a four-hour flight, the airplane lands and pulls up next to a fleet of buses. I descend the stairs in slow motion. I look magnificent. Somebody look at me! But there is no one to welcome us. We are shuttled, bused, and flown to the doorstep of every destination, escorted in through back doors under cover of police escorts and velvet ropes.

• • •

We get to the hotel a few minutes before 5 p.m. I grab the envelope that says JACKSON, NATE off the table near our private entrance. It has my room key, room number, another itinerary, and a room list. The room list comes in handy if you want to crank-call someone or desire to know the identity of your next-door neighbor whose head is less than two feet from yours while you both masturbate to free hotel pornography. Fifty-three men jerking: all in a row.

I arrive to find that the pay-per-view system's child lock is on. I can't find the smut in the guide. This happens sometimes. I just have to call down to the front desk and have them unlock it. No biggie.

—Uh, yes, hi. I'm not able to access the adult film selection in my room. I was wondering if you could remove the child lock.

—I'm sorry, sir, we don't carry adult films.

—Yes, well, good! That's good. Just making sure. Thank y—

Click.

But I'm not a pregame self-gratifier. I like to keep my weapon cocked and loaded. I believe that within my scrote swims an elixir that is a critical power source for my football performance, and to release it is to cripple my chances of finding glory the next day. But I still take advantage of the option. Watching pornography with no plan to discharge gives the film a depth that goes unseen when my intentions are lustful.

Dinner is a buffet of man food: chicken, steak, macaroni and cheese, spaghetti, some things called vegetables, fruit, soup, burgers, french fries, salad, etc. I walk down to the meal room, which is in the hotel's ballroom, with my playbook in my hand and fill up my plate with vegetables and pasta. I don't want to stuff myself too much. I need to save a little room for the late-night snack. All of the same options will remain, plus chicken wings and a dessert bar with pies, cookies, and ice cream. Piles of vanilla and chocolate ice cream are scooped, one after the other, into cavernous to-go boxes by a baffled hotel employee and drowned in syrups and sprinkles, then taken up to the room to be eaten in ecstatic solitude in front of an unchild-locked television.

But before the ice-cream social we have meetings: thirty minutes of position meetings, thirty minutes of special teams meetings, thirty minutes of offense/defense meetings, then a fifteen-minute team meeting. They are the compilation of the practice week's most heavily emphasized concepts and plays, already learned, already downloaded.

At our position meeting Pat returns the test that we turned in that morning in Denver. It's a written test that requires an expansive knowledge of the idiosyncrasies of the game plan from a tight end's perspective. I copied off of Mike Leach so I know I will have aced it. There are a lot of terminological obscurities related to the offensive line and defensive fronts and gaps and protections that I have never really learned. I know what to do on the field but I don't always know how to explain it in the terminology required of our pregame written test. So I cheat off of Mike. He always agrees but gives me a look I'm familiar with from high school, cursing me in his head for smoking pot under the bleachers while he was in the library at lunch studying.

Meetings, meetings, and a few more meetings: Watching film and going over plays on the night before the game always strikes me as pointless. If we don't know it now, we're not going to know it. But it's a video league. We are two-dimensional things.

Football has been subverted into a made-for-television event. Everything is so clear. Except it's not. The third dimension is what makes it real, violent, and dangerous. Consuming the product through a television screen, at a safe distance, dehumanizes the athlete and makes his pain unreal. The more you watch it, the less real it becomes, until the players are nothing more than pixelated video game characters to be bartered and traded.

After meetings a group of us sit around a table in the meal room eating chicken wings and ice cream. After I lick the spoon I ask Greek for some Ambien, which he produces from a small pouch and drops into my hand. I have no problem falling asleep without Ambien. I just like to take them and watch porn. At eleven fifteen, right when the drug is kicking in, there is a loud bang on my door that shakes me from my soporific boob-hallucination. It is Rich and Crime doing bed checks. Here! I gargle. That's all they need to hear, then they move on to my masturbating neighbor. If I hadn't replied, they would have entered my room with a master key and had a quick look around to make sure I was there and that I was alone. Little did they know that [Insert Popular Porn Star Here] is with me.

Soon I fall into a heavy, purple sleep and wake up in the morning with a powerful game-day buzz. I shower and put my suit back on and go downstairs for breakfast. We are on east coast time and it fucks with our internal clocks. I'll have some coffee. After breakfast, I get on the late bus (there are three different buses to the stadium and three different times in which to depart and indulge in whatever your away-game routine is) and look out the window at upstate New York. I listen to my music and prepare myself for combat. It's a beautiful day for football: 67 degrees and partly cloudy. It's the dawn of a new day, a new season, where anything can happen. We are Super Bowl bound, that much is for certain.

Tony is hurt so I am set to play a good deal on offense. I'm also on three out of the four special teams. One thing I love about playing special teams is that I am on the field for the first play of the game. It's the culmination of an entire week, an entire off-season of hype, and the electricity runs through the fans and through the grass and up through my feet, bringing me into a zone where I feel neither fear nor worry, only an extreme heightened awareness. I am light and strong, focused and lucid, and as I jog out onto the field, I can feel the energy inching toward a crescendo, toward the moment when the ball pops off the foot of the kicker and towns and cities and dreams full of potential football energy finally explode into the kinetic, into the now, forming a tidal wave that I surf through the circuit of the ultimate football matrix. There is no feeling that will ever replace that moment in my life. I know that now.

The opening kickoff of 2007 sails through the air. I chase it down the field, avoid the man who is supposed to block me, and tackle the returner at the twenty-five-yard line on the first play of our season. I'll be the Super Bowl *MVP*. The game moves along: tit for tat, nothing doing. It's a low-scoring affair going into the locker room at halftime. The Bills are up 7–6. Offensive coordinator Mike Heimerdinger, aka "Dinger," writes some plays up on the whiteboard. Based on what we have seen in the first half, these are the plays that he thinks will work in the second half. He goes over a few things, then I take a piss and jog back out onto the field.

The Bills kick off to start the second half. Wide receiver Domenik Hixon receives it and pushes up the field behind the wedge of offensive linemen holding hands and leading the charge. The wedge breaks down on the right side and Domenik bounces the kick outside the wedge, meeting Bills tight end Kevin Everett in a routine-looking football hit. But the result is not routine. Everett collapses to the ground and does not move. The crowd falls silent. Tony and I stand next to each other on the sideline.

—He looks dead.

We don't know how right we almost are. Everett has sustained a fracture and dislocation of his cervical spine. We'll later find out that his life was saved by fast thinking and perfectly executed medical treatment, which stabilized his spine and whisked him off to the hospital so we could finish our football game in peace, without the realities of what we were risking getting in the way. The show must go on.

It isn't long before the urgency of the game has erased the memory of what we have just witnessed. There is no time to consider the consequences. As the clock ticks down in the fourth quarter, we are down 14–12. But we are driving. With under a minute left and no timeouts, Jay pushes the ball down the field with precision, hitting Javon Walker several times and moving the ball inside Bills territory. We have Jason Elam waiting in the wings. He missed two field goals earlier in the game, very uncharacteristic of him. I'm the wing on the field goal team so I stand ready, too. If the clock isn't stopped we will have to run on the field and set up for the kick quickly. With eighteen seconds left Jay hits Javon for an 11-yard pass down to the Bills twenty-four-yard line but he can't get out of bounds.

—Toro! Toro! Toro!

The offense sprints off the field and the field goal unit sprints on. Tick-tick-tick. We can't see the clock but the Bills fans are kind enough to count down for us. Jason Elam doesn't have time to count out his steps. Our holder makes sure everyone is set and signals Mike Leach, who snaps the ball a tick before time expires. My man rushes hard off the edge and tries to jump between me and the end. I shove him in the chest and he twists through the air like a gymnast, landing on his back as the ball sails through the uprights for the win. Jason turns and sprints in the opposite direction, arms raised victoriously. I laugh and give chase. We need to celebrate this together. When I catch up I grab the back of his shoulder pads and tug him to the ground, sliding to a stop next to him just as everyone catches up and dogpiles us. It is a magical mountain of meat. We are 1-0 and Super Bowl bound.

There is nothing as satisfying in the NFL as going on the road and winning, because everything is going against you. You have to travel and stay in a new hotel. You have to face the unfamiliarity of a different city, a different locker room, different food, and a different time zone. The crowd is screaming, cursing your family, and laughing at your pain.

And you win anyway. Sixty-five thousand people fall silent. The locker room is jubilant.

A win means that all sins are forgiven, if only for a few days, and everyone can relax. Being able to actually relax in the NFL is rare. The pressures are too great. And they are constant. The head coach is under siege at all times, and it trickles down to everyone else. The industry hates losers. They ridicule them, defame them, and run them out of town. But we are winners. We are safe for the moment.

The back of the plane is boisterous. *This* is what the NFL is supposed to feel like. Grown men are happy and filling up red Dixie cups with vodka smuggled onto the aircraft. Booze in the empty postgame stomach of a football player is a bottle rocket. It hits the bloodstream dancing a jig and sings carols on the doorstep of the cerebellum. It feels like damn Christmas morning back here, in the working-class area of the Boeing 747, where FAA regulations are trumped by the laws of the jungle. We congregate in the back galley as the plane speeds down the runway and takes off into the sky, eight grown men leaning forward at a 30-degree angle like synchronized ski jumpers, heading back home to a proud city in love with its Denver Broncos.

We mingle in the back with the flight attendants. After all the trips, we have gotten to know them and they have gotten to know us. Not the faceless behemoths we are on television, but human beings with families and feelings. It is always the same group of us in the back of the plane. We are singing songs and telling jokes. As we cut through the night sky somewhere over Nebraska, John Lynch grabs the intercom microphone and flips it on.

—Excuse me. Is this thing on? May I have your attention, please, everybody. I'd just like to say, we live in the greatest country in America. Now please everyone, stand up, put your hand over your heart, and sing along with me.

He sings "God Bless America" at full tilt, and doesn't cheat his audience out of one single note. It echoes through the slumbering cabin and those of us in the back sing along, too, hands over our hearts, soaring over the Great Plains. This land is your land, this land is my land. Damn right we live in the greatest country in America.

After our laughter dies down, we all go back to our seats for a nap before the plane lands. An hour later we touch down in Denver and get on the buses, which appear unmoved from the previous day. The unloading of the plane and the bus ride take another hour. By the time we pull up to Broncos headquarters, we are all dead tired. And there is Bronco Betty.

Bronco Betty is the superest superfan in a world of superfans. She lives Bronco orange. She's at every charity function, every event, every training camp practice, and every game. She has a variety of health issues, and uses a walker, but she is unrelenting with her support for the team. And not just the team: the men. She knows everyone's name and everyone's story. She doesn't just watch the game, she *sees* it, and everything that happens. She sees every play that every player makes or doesn't make, and has a loving word for him regardless, win or lose.

Bronco Betty waits for us in a folding chair at the front gate of our facility to see us off as we leave for every road game and to greet us when we get home. Just Betty, alone in the Denver cold at three o'clock in the morning, sitting in her chair, happy as can be, festooned in her pin-accessorized Broncos gear and shouting personalized words of encouragement to each of us as we walk to our cars.

—I love you guys! Go, Broncos! I love you guys! I love you, Nate!

—I love you, too, Betty.

9

Rocky Mountain High
(2007)

The wake-up call doesn't wake me. I'm already up, lying on my back in my king-sized bed at the Inverness Hotel in Englewood, Colorado. My playbook is next to me on the bed. My clothes are laid out on a chair across the room. I get up, throw open the drapes, and behold the Rocky Mountains. Thus begins the same game-day ritual I've had since my first preseason game, more than four years earlier. Same room, in fact. And the same urgent feeling.

The security guard at the elevator sees me walking down the hall and presses the button to summon the lift.

—Good luck.

—Thanks, brother.

The elevator dings and spits me out into a buzzing lobby. Fans and family and friends in their finest orange and blue sip oversized coffees and laugh the carefree laugh of spectators. I walk past the reception desk, through a long hallway lined with meeting rooms, and into the largest banquet room at the hotel, which serves as our cafeteria.

—Morning, Chip.

—Morning, Nate.

Chip's our operations guy. He handles everything but the footballs: airplanes, buses, hotels, and meals.

I fill my plate with food to push around and sculpt. I'm not hungry. Eggs, potatoes, oatmeal, bacon, bagel, yogurt, fruit: a bite here, a bite there. That's all my stomach can handle. I napkin my plate and pick up a *Denver Post* that someone left on the table. "The keys to victory," "What to look for," "Key match-ups." Meh. I toss it back on the table, lean back, and sip my coffee. Several of the offensive linemen join me, fresh out of Mass, which is led by Bill Rader, our on-site spiritual advisor.

—Gentlemen.

—Hello, *Nate*.

The offensive linemen are the most devout Christians on the team. They attend Mass on game day and during the week. They attack their religious study as though it might offset the brutality with which they attack their jobs. Jesus levels them out. They're a thoughtful bunch: my favorite group with whom to break bread.

After a short chat about who watched what movie the night before, I leave and drive back to my house, shower, and change into my game-day clothes. My mother and father and Aunt Marsha and Uncle Bruce are in town for the game. The previous night, before I left for the team hotel, which even for home games we're remanded to, I showed my dad some plays in the playbook. I emphasized a specific goal-line play where I'm the only option to get the ball. Me or no one. In three years at Menlo College, I had 43 touchdowns. The plan was to keep right on catching touchdowns in the NFL, into the sunset. But it is my fifth year in the NFL and I have zero. Shanahan, God bless him, is trying to get me one.

We're playing the Jaguars in the third game of the season. I get home and say hello to my parents, who had the house to themselves last night. I take a shower and we board my Denali and embark downtown to [Insert Corporate Logo Here] Field.

—So it's a goal-line play you said to watch for?

—Yeah. I'll be on the wing on the left side of the formation. I go in double motion—actually it's like a triple-motion thing—and run a little flat route. The ball should be coming to me.

—All right, son. We'll be watching for it.

The play is designed to lull my man-to-man defender to sleep with a lazy double motion, then explode back down the line, behind the ass of the quarterback. He'll snap it just as I pass him, take three quick steps, and fire it to me at the front pylon for six points.

Driving up Interstate 25 from Greenwood Village to downtown Denver on game day is a meditative trip. The Rocky Mountains to the left, snowcapped with high-definition clarity, sparkle in the late-morning sun. I-25 is the main artery that connects Denver to its many southern suburbs. Among them, Dove Valley, which houses the Broncos headquarters; Greenwood Village, where I live; Cherry Hills, home of the rich and influential Denverites, including Coach Shanahan; and Highlands Ranch, the swingers capital of America (and where Charlie lives).

As we clear the suburbs and the city skyline comes into view, I-25 bends west for several miles and we pass Colorado Boulevard and University Avenue. The golden tower of Denver University's cathedral rolls by on the left, and somewhere out of sight on my right, Washington Park, the quintessential Denver patch: all sunshine and grass and dogs and volleyball and slip-'n'-slides and beer. Past the Whole Foods that sits perched above the freeway like a taunt. Past Broadway, Denver's oldest road (that's what my Denali salesman told me), which brings the swingers straight to downtown. Past the industrial warehouses off Santa Fe, where the road dips slightly then banks a hard right at an angle that's left the guardrail permanently scarred. After the road straightens out and ascends a small incline, the skyline once again comes into view to the northeast. And a moment later, like a flash, [Insert Corporate Logo Here] Field, straight ahead and slightly left.

Our stadium feels different than other stadiums. Where others are boxy, ours is rounded. Where others are rigid, ours flows. The upper rim of the stadium is not parallel to the ground but rises and falls like a hilly landscape, creating the illusion of moving water. And when all seventy-six thousand fans are in a frenzy, the field is a raft on a sound wave to Happy Town.

Normally I make the drive by myself. When the stadium comes into sight, my heart starts racing. I slow down for my exit and feel a stream of sweat run down my side. The Seventeenth Street exit wraps around and back underneath the freeway, past tailgaters and ticket scalpers, past the traffic cops and the early Broncos fans, all flooding into the stadium area with orange and blue everything. I show the security guard my player's parking pass and pull into the players' lot. Same time every time: 11:45 a.m.

But today I drop my family off at a restaurant downtown and take a different route into the stadium, no less festive. I park my car, grab my bag, and head for the locker room. To enter the bowels of the stadium from the players' lot, we have to walk down a long ramp that's lined on both sides by three-foot-tall iron dividers. Fans congregate behind the makeshift fences to watch us come to work in our civilian clothes, gazing upon us like ugly runway models. Some fans shout for autographs but most respect the game-day focus etched on our faces.

I pause at the locker room door and turn off my cell phone; I can be fined five thousand dollars if it rings on game day. On the other side of the double doors is Fred Fleming, our Everything Man, seated near the entrance.

—Cell phone off, Nate.

—Got it, Fred. Thanks.

The locker room is a large, open room: defense in the front, next to the equipment room, offense in the back near the showers. The carpet is blue with an orange and white Broncos logo in the middle.

I walk straight to my locker and swing my bag into it. Then I circle back to the equipment room near the entrance and grab a pair of gray shorts, change into them, and go to the hot tub next to the training room. Like every other week, here's Lou Green and Cecil Sapp.

—What up, fellas?

—What's up, Nate Jack?

Cecil has his headphones on so he can't hear me. He's reviewing his notes for the game. Running backs always have extra notes and tests and handouts they're studying. Lou's a linebacker. But we're all special teamers. And we came into the league around the same time and were on the practice squad together before being activated. We've done it the hard way.

After the hot tub I rinse off in the shower and walk back to my locker, change into my shorts and a T-shirt, put on my headphones, and take the field to warm up on my own before the team assembles. Those first steps onto the perfectly manicured grass, surrounded by seventy-six thousand seats, remind me of something I forget from time to time: I'm playing in the NFL. To be in the NFL, you can't be in awe of the NFL. You can't appreciate it while you're doing it. There's precious little time for self-reflection.

But as I jog onto the field at my own pace, the music in my ears, the stadium nearly empty, I take it all in. The crisp mountain air, the bright colors, the Broncos legends listed around the stadium in the Ring of Fame, and the faces of the fans who are given early access. The way they look at me. They don't know anything more about me than that I'm wearing cleats. That's enough.

After my warm-up routine, I reenter the locker room and sit back down in front of my locker. We have forty-five minutes before we'll take the field. I tape up my two big toes with two different kinds of tape, to stave off the blisters. I wrap a thin strip of tape around each finger and thumb between each knuckle. When I was a wide receiver, my fingers were pristine. I didn't need any finger tape. I barely needed gloves. But now I'm popping fingers all the time grappling with the beasts. The tape protects them.

Then I put on all my gear: socks, compression shorts, game pants, then my cleats. After tying up my shoes, I walk to the training room and jump up on Corey's table for my ankle spat. A spat is a tape job that wraps around the shoe and secures it to the foot, making the two feel like one. It is not as constrictive as a normal tape job, which goes directly on the skin, but still supplies noticeable ankle support. Even if there is a line for Corey and an empty table next to him in front of Greek or Trae, I wait for Corey. Corey spats me during the week, too. His spat feels like home. I'm not going to mess around with a new feel on game day.

Back at my locker, I pull on my shoulder pads and jersey. The pads are already fitted inside, and the whole setup is sitting on top of my locker when I arrive. I think about those doleful Sundays in the past when, after I had been declared inactive, Flip would come pull my jersey off of the shoulder pads I wouldn't be using.

I go into the bathroom to behold my body in its game-day armor. I'm not the only one who finds strength in the pre-kickoff mirror. There is a clamoring around the reflective glass. We are going on television. We want you to love us.

I'm impressed by myself. I look good and I feel good. My adrenaline is bubbling. I am a man built for hand-to-hand combat. The look is in my eye. The mirrored gaze. The dream fulfilled. I am a hungry animal. Time to eat.

Tight ends, running backs, receivers! Bring it up!

Bobby-T calls us up to take the field. We get in a line and walk out of the locker room, through the hallway, and out the tunnel. We stop at the back edge of the end zone and make a tight circle. D.G. puts his hand in the middle. We stack our hands on top of his.

—All right, fellas, we've put in the work, now it's time to have some fun. This group right here, everyone is looking at us to make plays. When you get that ball in your hand, make a fucking play. Make a *fucking play*! Forget everything else. Let's look out for each other out there, all right? And have some fucking fun. Broncos on three. One-two-three!

—Broncos!

And we're off, jogging down our sideline toward the opposite end zone, where we get loose. By this time the crowd has begun filling the stadium. The buzz is getting louder. We bounce around in the end zone and go through small drills with the tight ends: foot drills, ball drills, and some blocking to get the crank shaft ripping and the kinks out. Smack heads a few times, light off a few thunder bombs.

We break off with the quarterbacks and run routes, on the thirty-five-yard line, going in. Precise, crisp routes, exploding out of the break, snatching the ball that by now is as big as a blimp and squeezing with a strength reserved for emergencies, tucking the ball high and tight in my arms and exploding up the field into the end zone. I circle back around and toss the ball to Flip, who stands next to the quarterback and keeps him fed with balls. Then I'm back in line, waiting my turn again, nodding my head to the music that blasts from the speakers.

The horn blows. Riverside! Flip the line of scrimmage and take it down to the ten-yard line, coming out. We run set plays with the whole offense and the whole defense, thudding up but nothing too physical. A smack and a pop but let's save the ultraviolence for the game. After fifteen minutes of plays, the horn blows again and we jog off the field back into the locker room. As we make our way across the field, a cameraman lies on the ground filming us walk by him with a live feed to the Jumbotron. I walk past him slowly and look up at myself in giant form. Another quick reminder: you're in the NFL, buddy.

We have twenty minutes in the locker room before we come out to play. Quiet time. The calm. A pregame sheen of sweat covers me as I take my seat and again put on my headphones. The tunes are arranged down to the minute, to load, cock, and shoot my gun in unison with the first whistle. Seated on my locker chair, facing the large, open locker room, I scan the faces of my brothers. All of them are lost in thought, or lost in a thoughtlessness, a weightlessness that cannot be duplicated anywhere, ever—something that contrasts violently with the constant blah that follows an NFL career. This blah, when compared to the feeling right before stepping on the field, is what drives men to fits when they step off it for good. The nitro button is all the way down for too long and now the juice is all burned up.

—Praying in the shower, fellas! Praying in the shower.

The Christians gather in the large communal shower area, a secluded open space to kneel and pray. I joined them my first year until I realized I am closer to God if I sit still and listen to my music.

Coach Shanahan makes his way around the room and shakes every man's hand, tells him good luck, and gives him a pat on the back. He has an elephant's memory with the handshake thing and all things, really. Sometimes guys aren't in their seats when he makes his way around the room, so he goes around several times until he finishes the job, down to every last man. He never double-shakes.

Kickoff is at 2:15. The digital clock strikes 2:00 and Coach yells:

—Bring it up!

I throw my headphones in my locker and circle up tight around Coach, taking a knee and grabbing the hands of the men on either side of me.

—Okay, guys. We know what we have to do. We've had a great week of practice, but it's time to do it out there on the field. Leave it all out there. Do it for the guy next to you. Look after each other and execute, for sixty minutes, and we'll be just fine. Let's go get it done.

Then Coach also drops to a knee and grabs the hands of the men on either side of him. We are all locked in a chain.

—Our father, who art in heaven, hallowed be thy name. Thy Kingdom come, thy will be done, on earth as it is in heaven. Give us this day our daily bread. Forgive us our trespasses, as we forgive those who trespass against us. Lead us not into temptation, but deliver us from evil, for thine is the Kingdom, the Power, the Glory, forever and ever, amen.

Goddamn right amen.

The fireworks explode and we stampede through the mouth of the inflatable Broncos helmet and into the open air, lined on either side by shiny, smooth cheerleaders. Their hair bounces with an enthusiasm unfakeable and their skin glistens in the sunshine unmistakable. Game day in Denver is beautiful for many reasons. The cheerleaders are one of them. Their teeth are as white as the Rocky Mountain snow and their chaps hug perfectly the contours their job demands.

I run through their lane and hear the crinkling of their pompoms and smell their sweet perfumes mingling with the acrid smoke of pyrotechnics. I look skyward to the upper-deck seats, now fully packed and bubbling. Mile-high magic. The first few times I ran through the tunnel on game day, I was taken aback. We didn't have any cheerleaders at Menlo and our stands held five hundred people. Denver sells out every game—each and every game.

We take off our helmets for the national anthem and stand in a line to count the ways America is awesome. One, two, three . . . there are thirty-two cheerleaders in all. By the time I picture my life with each of them, the harmonized tune reaches its crescendo and the four fighter jets in formation come thrusting into view over the south end zone. The jets rip over the stadium ahead of their own sound, buzzing the white horse erected atop the Jumbotron in the south end zone and passing over the bubble of human energy, charging it further with the delayed roar of military turbo engines that buttress the final note of the anthem, carrying it into the distance with the jet fumes of America. Home of the brave. Whatever this is, it feels important.

• • •

Our captains take the field at the fifty-yard line for the coin toss. I don't pay attention. I'm pacing back and forth on the sidelines, smacking shoulder pads and head-butting. It doesn't matter who wins the toss. Heads or tails, I'm on the field for the first play.

—Kickoff!! Let's go! Kickoff team, bring it up!

We lost the toss. All ten of us, minus Paul Ernster, our kicker, who is kicking blades of grass off in the distance, bring it up tight around Scotty O'Brien, our special teams coach.

—Okay, guys. We know who they are. We know what they do. Watch their right return. And be ready for the double team on the fives. Keep your head up and pay attention to who is blocking you. They'll tip their hand. Avoid the block, stack him, and stay in your lane. Maintain it, fellas. And get down the fucking field! We're going deep middle, okay? Deep middle. Broncos on three: one-two-three!

—Broncos!

The huddle pops and we jog onto the field. Ten of us with one job: tackle the ball. The speakers are blasting our opening kickoff song by AC/DC, and the crowd is screaming. Not only have we waited all week for this moment, going through the rigmarole of practice and meetings, but so have the fans, going through the rigmarole of American life. This is for all of us.

I jog and take my place on the kickoff line and look up into the rafters. I see every face. I feel every breath. I'm the R3 on the kickoff team. The numbering system starts at the kicker. The first man to his right is the R5, then the R4, R3, R2, and R1. To his left, it's the L5 down to the L1. Paul counts off his steps, takes his three deep breaths, and puts his right hand up in the air. He pauses, drops his hand, and starts his approach. When he crosses my line of sight, I start mine. We all cross the thirty-yard line in lockstep with the pop of the ball off his foot and the explosion of the crowd.

And, zip. The portal closes over my head.

I hear myself breathing. And I hear my shoulder pads, plastic on plastic, echoing through the enclosed dome of my helmet. It's the hum of a high-performance vehicle on race day, purring in unison with the other nine muscle cars ripping down the runway, intending to flatten some silly rabbit. In our way are ten other vehicles: Ferraris and Mercedes and F-150s, diversified to throw off our assault. They have six Mercedes spread out directly in front of us, fifteen yards away.

They are the return team's first line of defense. As the R3, I keep my eye on the two directly in front of me. They're the most likely to try to block me. Every man on the kickoff return team has a specific blocking assignment, instructed to block a specific man. Some guys disguise who they're assigned to engage. But most don't.

As I run down the field no one is looking at me. The two men I'm watching dip inside. One of them goes to help his buddy on a double team of our R5, just like Scotty O said they would. The other sizes up and attempts to block Cecil, who is next to me. I have a clear path to the wedge, which is a group of three linemen holding hands and trying to clear a path for the returner. They're the F-150s: nearly one thousand pounds of meat coming downhill. Go ahead, try to stop them. It's your job, after all.

The R1 is the safety, the last line of defense, so he stays a few steps behind us. If the wedge comes free to me and the R2, and all the other guys get blocked, then the R2 and I must eat up the wedge and spill the returner outside into the arms of the R1. The returner has an upback, often a running back, who stands deep with him and leads him up through the wedge. But often, as is the case here and now, there is backside pressure from the kickoff team that forces the upback to bail out on the return and seal off the backside from pursuing coverage men.

The returner is on his own.

The R2 and I must take out three linemen and spill the returner outside and into the arms of our R1.

But Cecil easily avoids the block from the frontline guy and makes it down the field a step before me. He attacks the inside edge of the wedge and forces the linemen to honor his presence. The inside lineman jumps out to him. The outside lineman jumps out to the R2. That leaves me on the middle lineman and he's discombobulated and off balance and the wedge has collapsed. I give him a hard jab step inside and rip through with my inside arm, squirting into the hole and directly into the path of the returner.

I tackle him at the twenty-seven-yard line. First down Jags.

After a defensive stand near midfield, we get the ball back near our own twenty-yard line. A first-down pass loses two yards. Then I enter the game in our two-tight-end package called Tiger. Jay lifts his foot and I come across the formation in motion, settle, and start to lean. I've timed up the motions with Jay's cadence perfectly in practice. He usually snaps the ball as soon as I get to my spot.

But this time: no snap. Jay's looking at the defense, which is shifting and trying to confuse him. I'm leaning too much. I have to hold on with the tiniest proprioceptive muscles in my feet and hope he'll snap it before I take a step forward. Otherwise I'll get a flag.

Leaning, leaning, leaning.

—Set*hut*!

He snaps the ball and I run my corner route and turn to look for the ball. He throws it to the opposite side, to D.G., who ran the same route on the other side, and who catches the ball for a 34-yard gain before being pushed out of bounds. Yes! I look around for a flag.

Fuck.

—Illegal procedure, number eighty-one offense. Five-yard penalty. Repeat second down.

We punt two plays later.

I come to the sideline knowing I blew it, that I killed the whole team's momentum. We'd just stopped them on defense, forced a punt, and completed a 34-yard pass on second down. That's a great start. I ruined it single-handedly.

The first quarter ends scoreless. Broncos fans don't like scoreless. They're not used to it. They get restless when we don't score, even when it's a close game. They boo third-down incompletions. They want blood.

The first blood drawn in the game is by the Jags. They score a second quarter touchdown and line up for the kickoff. I'm on the front line, second man in from the right. Scotty O calls a middle return. The R3 is my man. The numbering system is the same as the opening kickoff except flipped. Looking at the kicker from where we stand, the R5 is directly to his right, the L5 to his left.

The more special teams I play, the more it slows down in my head. This allows me to be craftier. On kickoff return, that craft means not looking at the man I am blocking. After seeing the ball kicked, I turn and sprint back twenty-five yards at an angle, stacking the pursuit of the R4, the man inside the R3. I stare the R4 in the eyeballs so he thinks I'm coming for him. But I'm really looking at the R3 out of my periphery. (I wasn't always so crafty. I used to stare down my man the whole time, which meant he was prepared for the collision and tried to run through the back of my skull. I had to adapt, simply to save my brain.) Then, at the very last minute, with the R4 within steps of me, I plant my inside foot hard and explode past him and hit his buddy under the chin. He doesn't see me until it is too late. Now he's going to have to watch that on film tomorrow in front of all of his friends. Anyway, it was a touchback. No harm done.

We get the ball and start driving. A few first downs and we're into the red zone. Then we're down at the one-yard line.

—Jumbo! Jumbo!

That's our three-tight-end, two-running-backs package. I jog onto the field and take my place in the huddle.

—Okay, here we go.

Jay glances at me.

—Tight Left, Disco Motion, blah blah blah, on one, on one, ready, break!

I walk to my spot and get down in my three-point stance.

—Blue twenty-two.

He lifts his foot and sends me in motion. I rock back onto my feet lazily, trying to lull my man-to-man defender to sleep as I pass. Then I pivot and come back, equally lazily. The point is to make him think I'm doing a return motion and ending up in the same spot I started, so I can block a front-side run play. So don't look him in the eye. Make him think you aren't doing shit. Make him think you're blocking. Make him think you're *bored*. So much of offensive football is lying with your body, getting the defender to think you are going somewhere you aren't. Tell a story with your movements: a bloody lie! After my return motion, I turn on a dime and explode down the line of scrimmage back in the direction I was originally going.

—Blue twenty-two, set*hut*!

Jay snaps the ball quickly as I cross the ass of the guard. Stud cornerback Rashean Mathis is covering me. He catches on to the play and starts sprinting down the line of scrimmage behind his linebackers, who are hugged up on the asses of the defensive line. I'm sprinting, too. I feel slow: slow-footed, slow motion. The years have caught up to me. It's going to be close. Mathis has a good angle on me. Jay fires it down the line at the front pylon, which is my "aiming point." Either it will be a touchdown or an interception returned 99 yards for the Jaguars' own score. I reach out for the ball. So does Mathis. But he's one step late. The ball sinks into my fingertips and I squeeze it into my body.

Touchdown.

Five fucking years.

It's our first score of the game. The crowd goes wild—I think. The portal closes again and I can't hear anything. Total silence in my head. There were times over the years when the crowd was so loud I could hear my future children crying. But not this time. The kiddies are sound asleep. I spread my arms out like the wings of an airplane and soar around the back of the end zone smiling.

I circle back toward my teammates. D.G. comes to greet me first and we jump into the air and bump hips. I hold the ball tight in my right hand. Then I remember my family. By chance, I've scored in the end zone where they are sitting, a few rows back in the friends and family section, right along the goal line. I caught the ball on the pylon fifty feet in front of them. I look up to them and point. My father is smiling. My mother is crying. So is Aunt Marsha. Uncle Bruce, who has made the trip from Australia, is smiling and pointing at my mother saying, "That's his mom!" My parents have seen me score a million times in my life, on a million different fields and courts and swimming pools. Is this any different for them? Maybe. Is it different for me? Not really.

I walk to the sideline with the ball in my hand, absorbing the smacks and head slaps from my friends. They knew how bad I wanted that ball. They all want it just as bad. Some will get it. Most won't. I hand the ball to Flip.

—Nice catch, Nate.

He takes the ball and puts it into one of the trunks for safekeeping. He'll give it to me later. I take a swig of water. The extra point is good. I walk to the other thirty-yard line to take my place on the kickoff team.

The train rolls on.

After showering and changing, I throw my playbook in the playbook bin and fill up my bag with waters and Gatorades from the fridge. Then I walk through the double doors and out into the cavernous inner-stadium area where the buses pull in. Friends and family gather there to wait for us. Fresh off my first ever touchdown, I walk out and instantly see my father, then my mother, then Aunt Marsha and Uncle Bruce standing there as proud as can be.

We lost the game 23–14. But for a mother, the score doesn't matter so much. My mom has three criteria that she uses to judge a game. One, did I stay healthy? Two, was I happy with my performance? Three, did we win? Moms are ahead of the curve. The NFL is momless.

Up above our heads in the stands, the loss is all that matters. But down here outside the locker room, there is little to indicate the outcome of the game. Everyone's happy. If you are standing down here at all, you've already won.

It's here in the sea of friends and family that the humanity of my teammates is most apparent. A football team is thrown together from all corners of the country. Men are notified that they'll be joining some far-off team and then they go alone. They make their way alone. They do it by themselves, and in the process their individuality suffers. Their sense of self is lost. They become what they are expected to be—part of a unit. Yet for these fifteen minutes, after every home game, the true personality of each man can emerge, because here it's safe. I wasn't born on the fifty-yard line with a football in my hand. Neither were they. I'm not alone. We all are.

From the bowels of the stadium we walk back up the long ramp to the players' parking lot, passing the fans who line the steel blockades. There are five or ten faces on each NFL team that are known nationally. The big names get marketed heavily. The rest of us are nameless and faceless. But in Denver, the fans know all of us. As we walk past them after the game, they ask us by name to sign autographs.

Up close and personal with Denver's fans, it's no longer just me in my bubble of football solitude, teeth gnashing and clawing for some elusive gridiron glory. It's bigger than that, and it's right in front of my face: the smile of the eight-year-old girl as I sign her shirt. The look on the husband's face as he takes a picture of me signing his wife's chest, and the story he tells me about the fourteen-hour drive they've made for every home game for the last twenty-three years. The kid who produces a picture of me and him, a picture I don't even remember posing for, and now I'm signing it.

There's a genuine happiness spread across their faces as they meet their meaty unicorns. For a moment I can feel the horn in my forehead.

We leave the stadium and head south on I-25, back to the 'burbs for some very average Mexican food. Afterward, we go home and I tuck my family into bed. Win or lose, Sunday night's for partying. I need a release. I'm going to Spill.

Spill is in downtown Denver, a few miles from the stadium, made rich and famous by the athletes who get hammered there. It sits on the corner of Market and Fourteenth: a loud, narrow bar with brick walls and high ceilings. Flat-screen televisions mounted on the walls play videos that are pumped through the speaker system. The bartenders are friendly and pour 'em stiff and tall. The girls are plentiful and eager to please. Downtown is alive.

Every major professional sports team in Denver plays downtown. The Rockies play at Coors Field on Twentieth and Market, in the heart of Lower Downtown, or LoDo. The Nuggets and the Avalanche play at the Pepsi Center on Fourteenth and Auraria Parkway, steps from Spill. And we play a few miles away on the other side of I-25. This type of proximal action makes downtown Denver a happy destination. It doesn't matter, your preference. You'll find somewhere that suits you.

College drunks drink Coors Lite out of plastic cups in LoDo. Hipsters drink whiskey on Colfax and Broadway. Socialites stay south of LoDo near Spill in the three-block radius between Fourteenth and Seventeenth streets and between Blake and Wazee, drinking vodka and colorful shots. There are endless bars and restaurants, and endless girls with endless appetites for everything except food. To the unknowing, Denver seems to be a harmless, slightly boring mountain town in the middle of nowhere. Naw, son. Denver's a party town. Lucky me.

10

Watermelon Seeds
(2007–2008)

Okay, guys, have a seat. Listen, fellas. They're not doing anything we weren't expecting. Okay? It's pretty straightforward stuff. But you guys see how much they're overpursuing our keepers, right? We're going to go with our throwback keeper in the first series. Eddie, you good with that?

—Yeah.

I'm standing next to rookie wide receiver Eddie Royal on the north end of the locker room while Dinger holds court at the whiteboard. It's halftime. We are playing the Chargers in week five, a few weeks after my touchdown. I got my first start today.

But the start only sounds pretty on paper. Our opening play was a two-tight-end set. And Tony's still hurt. One play, then I was back off the field again and waiting. My main job is still special teams.

We kicked off on one of the last plays of the first half. At the end of the play, their special teams ace Kassim Osgood clipped me on the shoulder. I fell forward onto my knees, which were splayed wide. When I hit the ground I felt a tiny click in my left groin.

—Get yourselves warmed back up when you get back on the field. It's getting chilly out there. You don't want to get tight!

Scotty O is offering a bit of advice to which I'm not listening. I'm already warmed up. I feel great.

Last night after meetings I lined up for the needle again: 60 milligrams of Toradol, a powerful anti-inflammatory and painkiller. Ten or fifteen of us rely on it every game, physically and mentally. We live in pain during the week. We want to feel good on game day, and adrenaline isn't enough anymore.

I jog out of the tunnel, drink some sideline Gatorade, and line up for the second-half kickoff. The ball pops off the tee and all ten of us tear down the field. Twenty yards into my straight-line sprint, a sniper somewhere in the cheap seats pops one off and hits me square in the left groin, tearing the muscle off the bone with a yank that rattles off the iron of heaven's gate. *Thwap!*

(I've been electrocuted once. When I was a kid there was a Coke machine in the locker room at swim practice that was busted open and the wires were hanging out. One of the kids said that if you touch that red one to that blue one, something happens. No one would do it. I was soaking wet; armor, I thought. I touched them together: *thwap!* Frozen to a tuning fork of lightning bolts.)

The pop splays my left leg off to the side. I reach for my groin and hop on one foot. Total malfunction but I'm still on the field in the middle of the play. And I still have a job to do. The returner breaks a tackle and runs through the wedge, coming right at me. I have to make this tackle regardless. There are things on the line, after all: pride, glory, other stuff. Hopping along like that on the fifty-yard line, some things get put in perspective. One, Scotty O was right: I should have warmed back up. Two, I'm fucked. Three, somebody please tackle this man.

Just then a teammate trips him up and he comes to rest five yards in front of me. I turn and limp to the sideline, get dizzy, and almost faint as I reach the chalk. My vision blurs, tunnels, then steadies as Greek appears.

—What happened?

—Something popped.

—Where?

—Here.

I point to my groin.

—Okay, give it a minute and see how it feels.

—All right.

I turn and walk away gingerly. I'm done for.

I spend the rest of the game on the sideline with an ice bag stuck down my crotch. After the game I go home with instructions to return first thing in the morning. A pulled groin, they think. No biggie. My sister Carol and brother-in-law Jeff are in town for the game. We go out to dinner, a Brazilian meat-on-a-stick restaurant downtown called Rodizio. I limp and hop between tables and chairs. People look at me strangely. I am also meat-on-a-stick.

The next morning I continue the limp and the hop into the facility. My inner thigh and pubic region is very swollen. I'm walking timidly, slowly, straight-leggedly. It's ice and stim, right away. The stim, or "stimulation," uses electrical charges that pulse through two positive and two negative wire lines, like jumper cables, and are fixed to conductive pads and stuck on the skin, forming a picture frame around the injured area. The machine is turned on and the current surges diagonally through the meat, stimulating the healing process. The muscle jumps as the dial is adjusted and the electricity flows.

On top of the electrical pads, a bag of ice is fixed: formfitting and expertly tied. There is an art to tying a bag of ice. All of the air must be sucked out of the bag and the bag twisted and tied. And it can't be ice cubes. They don't conform to the contours of the skin. It needs to be ice pebbles, ice chips, ice dust, which will wrap around the injured area and freeze the muscle into a submissive postinjury state that will stop the swelling by slowing the blood flow, and, stay with me here, speed up the healing process.

The combination of an electric surge pulsing through the body, intended to stimulate blood flow, and an all-encompassing ice bag, intended to slow down that flow, might seem contradictory to someone in a position to consider it. But I am not in that position.

What's that, Nate? You want to rest it? Wait until it feels better before you start working it? Ha! You don't know anything about the human body. We have to speed up the healing process! Your body's natural reaction to the injury is incorrect. We can't trust it. We need to manipulate the body's natural healing process—shock it, change its temperature, strain it to the point of exhaustion, stretch it to the point of snapping, blast it with powerful anti-inflammatories and painkillers, then shock/temperature change again, strain/stretch again, pills again, and repeat. That's how you heal the human body, Nate. Make sense?

Another ice and stim treatment and then I drive to get an MRI. After filling out paperwork and wondering why I have to fill out paperwork, a nurse leads me into a back room, where I remove all metal from my pockets. Then she takes me into a large, all-white room and points to what looks like an alien pod. I kick off my slippers and maneuver slowly onto a plank. She ties my feet together and gives me headphones.

—Did you bring a CD to listen to?

No, I did not. The radio will have to do. She says goodbye as if I'm going on a deep-sea voyage and she leaves the room. She presses a button and the plank slides me into a magnetic cocoon. Her music preference query is a cute way of diverting my attention from the fact that I'll spend the next forty minutes in hell, as the hammers of Satan pound my skull into dust. Spiraled shock waves thunder down from Euclidean storm clouds, recording the fleshy echoes of my heartbeat. Bang after solitary bang: quicker, then slower, now faster than ever, now faster again. Finally, mercifully, the machine clicks off. I shudder as she rolls me out into the light.

—That wasn't so bad, was it?

I come back to the facility and roll on the training table for more ice and stim. The training room has a sterilized hospital smell to it: salves and creams and freshly laundered towels and cleaning supplies. I hate it, just because I know what smelling it means. I'm injured again.

Training tables in two rows line the back of the room and up front are the taping tables and a long counter with all the tools of the trade for trainers: tapes, goops, bandages, scissors, meds, rubber gloves, adhesive pads, X-Acto knives, nail clippers; you name it. The locker room is on the other side of the wall with the counter.

I lie on the table in the back corner sulking with an ice pack wedged in my taint, electrodes diagonally pumping through my testicles; electric chair for my spermatozoa, waiting for our team doctor to arrive and assess the situation. Once the MRI results are received, the doctor meets with the athletic trainer. They discuss the findings, come up with a course of treatment, then they brief the head coach. Once everyone is on the same page, they tell me. Make sense?

MRI findings:

High-grade complete tear and stripping of proximal left adductor longus and brevis, with distal retraction and about 5cm tear defect gap with intervening edema and hemorrhage. Strain of the adjacent pectineus and obturator externus and gracilis muscles and attachments.

Moderate proximal hamstring tendinosis and/or strain and scarring are seen, with longitudinal thinning and possible tearing of the proximal deep margins at the ischial tuberosity attachments bilaterally.

Dr. Schlegel, Boublik's partner, walks out of Greek's office and comes to my table with an excellent poker face.

—What's the word, Doc?

—Well, there are three muscles in the groin that come up and attach to the pelvis. The MRI showed that you tore two of them off the bone: the longus and the brevis. There's a five-centimeter retraction on those muscles, meaning they tore away and retracted from the bone by five centimeters. That's a significant retraction, but there is still one muscle intact and there are some fibers from the torn muscles that are connected to the pelvis, along with the intact muscle. In light of all that, we feel that we'll be able to avoid surgery.

—Okay.

—But this is a significant injury, Nate. We're going to give it a few days to let the swelling go down, and reassess. But we think we're going to send you to Vail for a procedure that we've had a lot of success with lately. Greek will tell you more about it, but it's an injection that uses your own blood to help speed up the healing process. It's called a platelet-rich plasma, or PRP, injection. It has proven to be a very effective treatment, especially with athletes. But this is at least an eight-to-ten-week recovery.

—All right. Thanks, Doc.

—No problem. And get some rest.

Doc Schlegel leaves and I don't. Greek comes out and asks me if I have any questions about what Dr. Schlegel has told me.

—No, not really. He said I might be going to Vail. And he said I'm done for two or three months. And he said to get some rest.

—Yep, exactly. Now let's get another round of ice and stim going.

Eight to ten weeks puts me into December. NFL teams have fifty-three active players. Coaches want all fifty-three men helping the team win: practicing, sitting in meetings, playing in games. If someone gets hurt, they have to decide how to handle it. Do we keep him on the roster and hope he gets better in time to play at the end of the season? Or do we put him on the injured reserve list, and call it a year? This frees up a roster spot and someone healthy can be signed and start practicing immediately. But injured reserve is a death sentence. Once you're on the list, your season is over.

The next morning I come into the training room on crutches with a watermelon cock. The swelling has overtaken my genitalia, mushing it into the opposite corner of my crotch. Black-and-blue ribbons curl up and down my inner thigh. A mile's worth of chains are wrapped around my pelvis, intertwined, cinched tight and locked. I am worthless. The medical staff reaches the same conclusion: no point rushing this injury back on the field. Greek tells me I am going on injured reserve. My season is over.

What do you call a football player who's not playing football? You don't.

On Wednesday morning we watch special teams film and I see my injury on the big screen. There I am running down the field. But it doesn't feel like I'm watching me. I'm watching someone else: something else. I'm watching a video game. When the video game player hitches and starts hopping, I think, what the fuck is wrong with that guy? Fucking run! Run, you idiot!

I crutch in and out of a few meetings that day but soon realize there's no reason to be there. I'm not going to play; no need preparing like it. Besides, a cripple hobbling in and out of meetings every day is not a welcome sight for those trying to focus on the task at hand. A cripple makes them confront the obvious: you're one play away from ending up like me. NFL football players must believe they're invincible or they'll get trampled on the field.

As I lie incapacitated on the back table on Wednesday, one by one my coaches come to pay their final respects. They tell me to hang in there; work hard on my rehab and I'll be as good as new. They use hushed voices and somber tones. They wear grave looks on their faces. This is my eulogy. They are moving on in order to save themselves. They have no choice. I'm dead until April.

The limp and the hop; perfected in silent corridors, echoing off the tiles of an empty shower room. The watermelon ripens in the corner of my manhood. After four or five days, the swelling has gone down enough to go to Vail for the PRP injection. In the middle of the week, Sara flies to town and drives us up into the mountains while I squirm in my seat like a thing unhinged.

The procedure is set for the morning. We go up the night before and check into the Sonnenalp resort, dimed by the Broncos. It is just down the street from the highly esteemed Steadman-Hawkins Clinic, in the heart of beautiful Vail. It is a great place to convalesce. I haven't seen Sara for a while. And it's nice to get out of my house in Denver. The room at the Sonnenalp is nice, too. It has a mountain lodge feel: spacious and cozy. We have a fireplace and a big bathtub and despite my disability, I am determined to take advantage of the accommodations.

As I lie on the bed, and as she scuttles around me rearranging things, I grab her and pull her on top of me. I kiss her and reach between her legs. She protests the advance, if not simply to spare me the embarrassment. But I persist. She stirs passions in me that even a watermelon cannot deflate. She is the most beautiful creature I know. And my subconscious knowledge that we won't last makes certain I'll appreciate her while we do.

Delicately we remove our clothing. She maneuvers herself on top of me, flinching only slightly when she sees the rotten fruit. But her touch sparks a miraculous blood flow. In defiance of the forces that oppose it, my valiant soldier rises to salute his muse. In appreciation of the gesture, she gently accepts.

Slowly we squirm, but I am wilting. I remind myself of the objective—coming—and try to clear my head. The misery of life's worst moments, deepest pains, and saddest bedridden days of depression are no match for an orgasm. But the body on body, the thrust and the cushion, the sensual bumping and pressing that accounts for the real action of love, is nearly impossible.

But my will leads the way, and we find the elusive rhythm. Her skin flushes, her hair follicles dome, and her lips redden. I feel a spark ignite in the depths of her ocean. Faint, then less faint, creeping toward us cautiously, the tide unlocks the gates of her hibernat-

ing libido. I paddle out to meet her. In slow motion and clothed in
windblown linens, we splash into each other's arms as the symphony approaches its climax. Just when the final note is to be carried
into eternity, the conductor drops his baton, the instruments crash
to the ground, and a solitary oboe pushes out one flitting note. Poof.

—Did you?

—Uh . . . I think so.

Watermelon seeds.

The next morning I crutch into the Steadman-Hawkins Clinic for
my PRP injection. They give me a hospital gown and I crawl onto
a gurney. Of course my nurse is hot. This pattern develops around
every genital area injury I have in my career. The more emasculat-
ing and uncomfortable the injury is, the more attractive the woman
will be who treats it.

She ties up my arm and pushes in the needle. They need a good
deal of blood for the procedure. One vial, two vials, three vials, four:
I lose count. Many vials later, she pulls the needle out from under a
cotton swab, presses down, and covers it quickly with a Scooby-Doo
bandage.

She leaves the room with the vials and comes back with my
blood in a bag. She opens a large circular machine and fixes the
bag inside, closes the top, and turns it on. It is a centrifuge, and as
it hums and spins, the properties of the blood separate into smaller
pouches on the sides of the machine.

—See that one there? The one that looks like urine? We don't
need that. But see the dark, thick red stuff? That's the good stuff.
Look at that. That's beautiful.

She fingers the bag.

—That's the platelet-rich plasma. That's what's going back inside you.

She points to my balls.

After thirty minutes, the machine clicks off, the separation complete. She takes all of the bags and leaves me on my gurney to count the holes in the particleboard ceiling squares, wondering if I could pop one of them off and find the ventilation ducts like John McClane in *Die Hard*.

Shoot. Ze. Glass.

After a few hours of waiting, it's time for the injection itself. They'll sedate me, my nurse says, because of the location of the injury. It will be painful and squeamish otherwise. For both of us, I assume. They wheel me into an operating room and I look around frightened. All of these people in masks. Why so many people? And why the masks? I'm not wearing a mask! Where's my mask? And it's so cold in here; so cold.

The anesthesiologist introduces himself and pushes a needle into my hand in one motion. My nurse pulls up my gown and swabs my groin with alcohol. I wonder if she can see my penis. Nurses have kind eyes. I feel the drugs hit my bloodstream, tubing through my veins and arteries. At the same time I feel a trickle of alcohol catch momentum, run down a ridge, and hit the tree line in the crease between my leg and my crotch. The race is on. My eyesight fogs over. My lips feel big. It isn't cold anymore, except for the river below, raging toward a protected marsh. My nurse watches over me maternally. I close my eyes and surrender to the drugs.

In the hallway on the way to the recovery room, I meet Dr. Marc Philippon: PRP injector, world-renowned hip surgeon, excellent human. He is still masked and wears a full blue hospital suit. His floppy Scandinavian blond hair, flowing from underneath his light blue skullcap, and his bright blue eyes strike a vibrant image in my loopy mind.

—Everything went great, Nate: really, really great. You'll be fine in three weeks. Now rest up. All you can do is rest.

I manage a single question before his wingtips are echoing off the linoleum floor of the long hospital hallway, blond hair bouncing to the beat.

—Will you tell that to Greek?

• • •

I am back at work the next morning; strapped again to my electric chair. My only job now is rehab.

At 8 a.m. I jump on my table.

At 12 noon I jump off and go home.

The four hours in between vary slightly from day to day, but closely follow the protocol for "torn groin muscle," ramping up the work slightly every day, depending on how my body responds.

The injected plasma encases the damaged tendons and hugs them with nutrients, forming a bridge of goop that the retracted muscles can cross before reuniting with the bone that held them since the womb. I sit on the table and meditate through the hours of ice and stim, picturing the PRP as a fleet of noble warriors sent to save a town from a bloodthirsty regime; sort of like the Three Amigos. My torn groin is El Guapo.

Every day I go home at noon and sit on the couch and think. That's it. Sit and think, and find your way to God.

Some days I have bright ideas for home improvement. I live in a big house alone. It's my house! I bought it! I go to Home Depot and buy every color of spray paint they have and get to work on a mural in my master bathroom. Stencils and acrylics and oils and brushes follow. The walls are my canvas. And my targets.

The next day I throw butcher knives at the walls. The one after that I string pinecones from the rafters to usher in the snowy season. I read the first thirty pages of lots of books. But I can't concentrate. I take guitar lessons once a week in Boulder from a bluegrass guitarist named Brad. That gets me out of the house and gives me something productive to do. I write a few songs, a few poems. I keep a journal that drips blood when I open it.

The entertainment center in my living room is set into the wall about three feet deep and has a six-inch-thick drywalled shelf, skeletoned with two-by-fours, that's built into it, dividing the top from the bottom. Sometimes I have the TV on the bottom, sometimes I move it to the top. But I really want it in the middle. That fucking shelf is no good. But it's built into the damn house!

One day after rehab, I go to Home Depot and buy everything I need to demolish it: multiple power saws, hacksaw, crowbar, sledgehammer, and disc sander. Then I go home and tear my living room to shreds. When I'm finished, there is a layer of dust caked to the furniture. But the television now sits proudly in the right spot, surrounded by a torn-up wall with cables hanging out of it and exposed two-by-fours and drywall.

The injury also gives me more time to spend at home tracking the movement of the family of mice that has moved in. One night, after ingesting some unprescribed herbal medication, I am cleaning up in the kitchen. There is a single light on overhead. I lift a pot and a mouse darts across the countertop. I jump out of my slippers, hit my head on the ceiling and moonwalk into the pantry. While I'm there I inspect the area and find a collection of turds in the corner. I'm not surprised, as I often poop there. But in the opposite corner of the pantry is a collection of much smaller turds: mice! The next day at rehab I can hardly contain myself. I relay the story of the mouse to everyone in the training room. When I'm done with my workout I go to Home Depot. By now they know me.

—How did the demolition go?

—Eh, ya know. Least of my worries now. I have mice.

—Mice?

His face lights up.

—Follow me.

He takes me to an aisle that's lined floor to ceiling with widdle fuzzy murder tools: snap traps, glue traps, trapdoor traps, poison pellets, poison goop, poison juice, poison poison. I decide on the old-school, hinged snap trap. It was not a large mouse. Yep, this should do it. I pay for my tools and race home. I'm so excited I barely notice my depression.

I set one trap next to the small turd pile and the rest in and around the kitchen. Then I leave the house to summon the death angel. When I walk in the front door a few hours later, a cold wind blows through me. A rodent lies dead on the kitchen floor. He died of a broken heart/neck. I lay another trap in the same place, knowing his bride will come to pay her last respects. When I wake up the next morning, the trap is gone. I find it underneath the dishwasher. Bitch got snapped and dragged herself under the dishwasher, where she wiggled free and was off in the night. Well played, Minnie, well played.

From that day on I sit on the kitchen counter every night with an airsoft gun and night-vision goggles, waiting to see if there are any more. I have soaked the airsoft pellets in poison juice. I'm not going to be made a fool of by some common field mouse. Eventually I get all five of them and seal up their entry points, which I find on either corner of the garage door. I feel triumphant in my kingdom of solitude.

But I am a living shadow outside my house, ducking in and out so as not to be seen, not to have to speak to anyone. At the facility I am the model of hard work. I earn employee of the month honors twice in a row. Pro Bowl center Tom Nalen and I were injured in the same game and were put on injured reserve on the same day. Our daily lives at the facility are mirrors. But it is his fourteenth season in the NFL. He is the glue that holds together the offensive line, which holds together the run game, which holds together the offense. Without Tom's presence, we struggle to find the rhythm all year long.

One solace I find in my injury is the opportunity it has afforded my friend, fellow tight end Chad Mustard. He played for us the previous season. He is a six-foot-six former basketball-playing beast with quick feet and a big heart. He is a huge man, so huge that when there was a shortage of offensive tackles, they gave Chad a new number and stuck him next to the guard. Go for it, Chad. He does anything that is asked of him on the football field. And he can do almost anything. The more you can do, the more they ask you to do.

But when we signed Daniel Graham in the off-season, Chad got squeezed out. He moved back to Nebraska with his wife and started substitute teaching. When I got hurt, they called Chad and he was on a plane back to Denver. He even got his old locker back, right next to Tony and S.A.

We go 7-9 and miss the playoffs for the second year in a row. It is our first losing record since 1999. I receive a clean bill of health after the last game. In the off-season I become an unrestricted free agent for the first time in my career. That means I can test the market if I want to. But Ryan says the Broncos want to re-sign me. Coach Shanahan has been very good to me over the years. There's no reason to leave. Plus, chances are there wouldn't be many takers. I am an undersized tight end with body issues. The fact that they want to give me a new deal in spite of my recent injury is a great sign and not one that we should overlook.

We begin negotiations on the first day of free agency. What that means is that Ryan and Bruce begin negotiations with Ted and his team. Ryan wants to get me guaranteed money up front in the form of a signing bonus and add a few conceivable escalator clauses in case I end up catching a shitload of passes and touchdowns. I say, go for it, buddy! Get me all you can get me. My life hurts. But pain is glory. And glory is money.

Ryan keeps me updated. They are negotiating this, they are discussing that. They are considering this, they are evaluating that. Eventually he calls me and says that he thinks we have a deal. It is a two-year deal, league minimum for sixth- and seventh-year players, plus a $425,000 signing bonus, a roster bonus, a workout bonus, and a handful of escalators in the event of a breakout season. This is amazing, I say. I'm mainly talking about the $425,000 up front. Guaranteed money is all that matters in the NFL these days. As we all know by now, no one is ever safe. I can sign a five-year $50 million contract, then get cut the next day and never see a penny: unless I get some up front. Guaranteed money is money already earned. I tell him, yes, yes, and yes. Let's do this.

Two weeks later, Ted gets the tap on the shoulder and is fired as general manager. It's not just the players who sleep with one eye open: it's everyone in the building.

The week before we report back for training, I take a trip to Los Angeles to play in the Playboy Golf Tournament, where half-naked chicks doing keg stands is a requirement. Before Jake retired, he plugged me in with the Playboy people. Now I get invites to the events I used to sneak into. There are six to eight girls at each tee box. Early in the day, they are predictably unenthusiastic, lounging under trees and texting. But as the booze starts flowing and empty promises are shot through the well-rehearsed O's of cigar smoke, their spirits lighten and the fun ensues. Each tee box presents a new group of barefoot drunk chicks doing cartwheels.

The actual Playboy Playmates are not tee-box girls. They're past all that. Instead they drive around in golf carts and pose for pictures with the foursomes. Over the years, I've enjoyed the company of many a *Playboy* mag. I made friends with the pages. They have kept me company through sleepless nights. And so it is with a surreal kick in the scrote that I bury a ten-foot putt, pull my ball out of the cup, and glance up to see a familiar silhouette walking toward me. The sunlight holds the outline of her black hair and sheer white dress, which clings tightly to her hourglass frame, keeping no time but eternity. She stops directly eclipsing the sun and turns to look behind her. As she does, the light shoots through a violet gemstone hanging from her neck. Wait, I know that stone! It is Hiromi, my travel companion from Germany. She is wearing the amethyst that I carried in my jetlagged hallucinations, through the doors of perception. I don't tell her that we've known each other for years. I don't have to.

11

The Last Dislocation
(2008)

After sitting out most of the previous season, I'm eager to get back to work and back into the violence of the trade. I want to prove that I deserve my new contract. Enduring the pain and violence makes me feel like I'm earning my money. Violence is football's winning formula. The game depends on it, and you can't go through the motions in practice expecting to be in murder mode when the season starts. A taste for blood comes during training camp, when it's hot and grimy and players are fighting for their lives.

But as the visceral reality of constant violence digs into me, I begin to question everything about my life, wondering how I got myself into this mess and if it's too late to get out. Worse still, Rod Smith is gone now.

After playing the entire 2006 season, Rod had microfracture hip surgery in the off-season. The surgery did not work. His hip was ripped to shreds from the years of football. He spent all of 2007 on the sideline in the hopes of returning. But his body would not let him. He had hip replacement surgery at the end of the season and seven months later, as we arrived to camp, he announced his retirement. My role model has left the building. So have most of my close friends. I can't be too far behind.

Every morning when I pull myself out of bed, *North Dallas Forty*–style, I play out the conversation in my head, what exactly I will say when I go up to Coach Shanahan's office and tell him I'm quitting, that I can't take it anymore. But by the time I pull into the parking lot, I have once again convinced myself that I am a warrior, and this is my war.

Training camp is an attack on the mind: an attack on one's sanity. Enduring it for six years has desensitized me to pain and anguish. Pain isn't rigid. It's a choice, a weakness of the mind, a glitch in the system that can be overridden by stones and moxie. I find my switch and flip it. People often asked me how bad it hurt to get hit by those huge dudes. The truth is that it doesn't hurt at all. The switch is on. I can't feel a thing. My body is a machine and my emotions are dead.

But the years of abuse are taking their toll. Misaligned joints, stretched ligaments, bruised bones, overworked muscles, and a jangled brain keep pace with an ambitious football mind. One play at a time: one day at a time. My football mind is stronger than my human body.

After morning practice we have a few hours to ourselves. We have everything we need at the facility: inflatable mattresses for naps, video games, and DVDs on several large TV screens in the players' lounge, a daily movie screening in the team meeting room, and plenty of boxes of tissues to sit in a dark corner and cry with.

Because of the evil dream clown that has been tormenting my sleep since San Francisco, I don't like to fall asleep between practices. Instead I sit in the locker room and shoot the shit with Domonique Foxworth and Hamza Abdullah and Brandon Marshall. I'm learning to play acoustic guitar. I sit on the floor and strum the only three chords I know. If someone walks through the locker room we make up a song about him. It's meant to humiliate and cut deeply, in the hopes of unearthing a crippling insecurity. The more distraught our victim, the more aggressively we laugh at him. The longer he stays, the worse it gets, until he finally realizes he is dealing with madmen who have lost the ability to empathize, and he scurries off. I'm not concerned about another man's feelings. I don't even have time for my own. This follows me off the field and out into the world, where people's concerns seem weak and pointless. Pain is a choice.

I don't realize it at the time, but the ability to relax and be an asshole between practices is a product of becoming a seasoned pro. My early years in the league were fraught with nervous tension. I was in no mood to joke around. How could I? I was on my deathbed. But as the years have gone by, conquering the daily struggle has become ingrained in my psyche. My mind is finally free inside the machine, sort of. I exercise the freedom in questionable ways.

We have ten or so ball boys who spend training camp with us, helping out Flip and his staff, who are bogged down with work. They range in ages from ten to eighteen and are typically sons or cousins or friends of someone in the organization. By the end of camp we are all great pals, and parting is sweet sorrow. One afternoon I bump into one of our high school ball boys in the hallway. He is texting on his phone.

—Who are you texting?

—My girlfriend.

—Let me see a picture.

He hesitates.

—C'mon!

—Okay.

He clicks through his phone.

—Here.

—Nice. Does she have a sister?

—No . . . but her mom is hot. And she's single.

—What? How old is she?

—I don't know. Thirty-something.

—Hook it up, Chad.

—Really?

—Yes, fucking really.

—Okay, I'll try.

—Don't try, Chad. Do it!

He rolls his eyes.

—Chad! I'm serious.

—Okay! Jeez.

I let him out of the headlock and walk away.

Seen from space, training camp is hundreds of men seques-tered in squalid conditions for a month with no female presence anywhere, except for the women who have come to watch practice. As camp wears on I spend more and more time scanning the crowd for good-looking girls whom I can point out to my equally deprived teammates.

—Look, forty-yard line, white shorts. Do you see her?

—I think that's a guy.

—No way. Look at her ponytail.

—Look at her mustache.

It's a common phenomenon. It's what happened when beautiful Inés Sainz walked into the Jets locker room in 2010. They hadn't seen a woman in weeks, let alone a beauty queen with a Spanish accent and an outfit that clung to her frame like the scent to a rose. Their reaction, the reaction that sent the *media*, not Inés, into a frenzy, was not rude or aggressive. It was boyish and harmless. It was campdick. Campdick led me back to Chad the next day, pes-tering him about the mother of his girlfriend.

—What's the word, dude? Tell me something good.

—She said you can call her.

—What?! Yes, Chad! Yes! Okay, what does she look like? Can you get me a picture? I need to know what she looks like, dude. I'd love to take your word and all but you understand. Have your girl-friend send you a picture.

—Okay, okay. God, you're weird.

—Just do it.

Later that night, Chad approaches me with his phone in his hand and a look of accomplishment on his face.

—I got it!

He shows me the picture. It's blurry but he is right. She's hot. This is exactly the excitement I need: a nonfootball distraction to remind me that I'm not a football robot.

—Perfect. Give her my number. Nice work, you little *fucker*!

The next day she texts me and I text back. Back and forth we go. She implores me to send her a picture of me since I have seen a picture of her. I'm considering it over a bowl of Cheerios when one of the ball boys, named Aaron, son of superlawyer Harvey Steinberg, sits down at my table at the cafeteria with a solemn look on his face.

—What? I ask him.

He looks at me sheepishly.

—What is it, Aaron? Out with it!

He then reluctantly tells me that Chad and Bobby, another of the ball boys, are playing a trick on me. That isn't Chad's girlfriend's mom's number. That's Bobby's phone. I have been texting *him*.

—Well, what about the picture?

—They got it off the Internet.

—What? The *Internet*? You're lying! Don't lie to me, Aaron!

—I'm not lying. I promise.

I've been had, duped, played like the delusional campdicked loser I am. I text *Megan* and tell her she is going to fucking pay and to be ready for the fires of hell to engulf her sneaky, conniving little face. The next day I recruit my best friends on the team for some retribution. My plan is to get an army of dudes to subdue them both in the locker room, wrap them up like Siamese mummies with athletic tape, dump every available noncorrosive liquid and powder on top of them, and throw them incapacitated into a freezing cold shower, where they will squirm around on the tile floor until they either wiggle out or some sympathetic pussy sets them free: a fairly standard hazing. The problem is, in order to recruit members for this special-ops mission, I have to tell the story. I have to explain to them why they should commit this act of torture on these pubescent teens. The reaction I get is not what I had hoped for. Everyone thinks it's hilarious and declares them geniuses.

—Well, fuck you guys, then! I'll do it on my own.

But I can't do it on my own. Thankfully, Tony Scheffler and our new linebacker, Niko Koutouvides, see how desperate I am and help me corral them as they walk through the locker room collecting laundry bags that evening. But Tony's and Niko's hearts aren't in it. My heart, though, is fully invested. I do my best to execute my plan on my own. I get them tied up, sort of, and pour ketchup and mustard and baby powder on them as everyone looks on and laughs, presumably at me and the frantic pace with which I'm trying to even the score in a game that has already been won. Sweating profusely and out of breath, I drag them into the showers, turn on the cold, and leave them to fend for themselves. It doesn't take long before some sympathetic loser is cutting them free.

Breaking up the monotony of camp is the arrival of the Cowboys for three days of practice leading up to our preseason contest against them at [Insert Corporate Logo Here] Field. Practicing against another team is difficult. There is an etiquette and tempo that exists at practice that varies slightly from team to team. For us it goes like this: unless it is a live goal-line drill, which only happens maybe once a week in camp, no one blocks below the legs or tackles the legs or touches anyone's legs in any way. Everyone stays off of each other's legs because that's how people blow out ACLs and break ankles. You also don't take anyone to the ground. You simply "thud up" with a nice pop. Coaches see if you are in a position to make a tackle. No point putting an exclamation point on it. You also don't block a guy in the back or pull on someone's jersey from behind or take a kill shot on a player or dive on a pile or touch the quarterback *at all*. You protect each other if possible because the sixteen-game regular season is brutal enough as it is. And after all, we're all on the same team.

But when another team comes in to practice, the lines get blurred. The previous year we had gone to hot, muggy Dallas. We had come directly from our preseason game in San Francisco. We were tired and nursing injuries and didn't perform well. They were hooting and hollering. They wanted us to know who they thought they were.

After the few days of practices we played the game and got our ass kicked. And they broke a few unspoken rules in the process. Preseason is a time to run your basic shit. No one game-plans heavily for their opponent. But the Cowboys had blitzes and trick plays and all kinds of nonsense. Then they ran up the score when they had us whupped, airing it out with only a few minutes left in the game and a hefty lead.

In the team meeting the night before they arrive in Denver, Coach Shanahan reminds us of what happened the previous year. We remember. When practice begins the tension is thick. Wade Phillips, the Cowboys' coach, has surely just given them the same speech. There is shit-talking, big hitting, and a lot of unnecessary celebrating. HBO's *Hard Knocks* is following the Cowboys for their annual training camp series, which does little to limit the theatrics. I stand on the sideline laughing about it with one of my old 49ers teammates, Terrell Owens. He's a Cowboy now. I haven't seen him in years. As we talk I recall a chat we had six years earlier at the lunch table in San Francisco. I was a baby deer slipping on the training camp ice. He was a superstar. I wanted what he had. I was trying to pick his brain. But it was already picked clean by the machine. He told me then that basketball was his real passion.

—This shit, football, is for the birds.

Football is for the birds. Here we are six years later: a couple of fucking peacocks.

Then I have to get in the huddle and social time is over. There are two full-sized grass fields at our facility, as well as an almost-full-sized field-turf field. One grass field is our offense versus Dallas's defense. The other grass field is our defense versus the Cowboys' offense. It stays that way all practice, except for special teams, when we all gather on one field. There are hundreds of people separating the two fields.

The Cowboys' defensive line is enormous and freaky strong. By now I know what I have to do to block a beast like that: crack him in his jaw with the crown of my helmet, then grab him tight and hold on. And still I'm at a disadvantage. Every run play is internal chaos and insecurity pushed through a makeshift "get-it-done" filter. Visualize the task and fucking do it. On one particular play I'm playing fullback, as tight ends are often asked to do. Years ago Mike Leach had adopted the "the more you can do, the better off you'll be" credo, and it worked. I've followed his lead and it has kept me around, too. But among the do-everything guys, we have another credo: "the more you can do, the more they make you do." That's what happened with Chad Mustard, too, when they moved him to tackle. In some cases—such as this one—it works against your better interests. I hate getting put in at fullback. Whenever I do, I look at Pat McPherson like, Really? He looks at me like: Yeah, really, motherfucker! I came into the league thinking I was going to be a Pro Bowl wide receiver. Now here I am in a three-point stance lead blocking through the two-hole, about to get my dick ripped off.

The ball is snapped. I take a slight sidestep with my left foot and come downhill through the hole. I am off balance as I bend to the right to find my man, middle linebacker Bobby Carpenter. We meet with a hearty pop. I take it all on the point of my right shoulder, feeling it buckle and separate. A power hose of fire ants shoots through the right side of my body. My arm hangs dead like an ivory-scalped elephant trunk. But a shoulder separation is only pain. So deal with it. After practice it's examined.

Brandon Bushnell, M.D.:

REASON FOR VISIT: Right shoulder injury

HISTORY: During today's practice with the Cowboys, Nate was tackled and came down hard on the point of his right shoulder [*sic*]. He experienced immediate onset of pain. He did not feel like anything popped out of joint. He was able to complete practice. No tingling and numbness in his arm. No other complaints. Soreness localizes to the anterolateral aspect of his shoulder and is worse with attempting to raise his arm above his head or across his body.

RADIOGRAPHS: X-rays reveal anatomically aligned AC joint. No evidence of fracture, dislocation or subluxation.

ASSESSMENT: Right shoulder grade I AC sprain.

PLAN: After discussion of the risks and benefits including drug reaction, infection, skin color changes and other issues, the patient wants to proceed with an injection of the AC joint. Under strict sterile conditions, Dr. Schlegel injected the AC joint today. He tolerated the procedure well with no complications. Within a few minutes after the injection, he did report increased relief of his symptoms. We are going to have him work with the training staff on ice, stretching, and other modalities. He will be cleared for play and we will have him wear a doughnut pad over the AC joint. We will continue to follow him for improvement.

Hello again, Mr. Needle, how well you mask my pain. But will I pay for this quick fix some other lonely day?

The needle works and I forget about my shoulder. Soon there will be new pains to address. It seems as if my mind never takes on more pain than it can handle. An instability or budding injury might be waiting in the wings, might already be symptomatic, but it's always polite enough to wait until the previous injury has significantly improved before roaring to life.

In the locker room after practice the next day, our defense is in a frenzy. There has been a fight on their field. It had started with

Cowboys offensive tackle Flozell Adams and our linebacker, Nate Webster. Nate is a small linebacker and Flozell is a giant. He is the largest, scariest man I've ever seen in person. But Nate isn't one to back down.

We watch the fight on film that night. It starts like most NFL fights. The whistle blows but the players don't stop. Something snaps. Fuck him. Fuck him. Then an explosion of fists flying from a frenetic jumble of padded bodies: everyone swinging for the fences. Flozell ended up with a Bronco helmet in his hand and was swinging for the fences, too. Thankfully he didn't hit anyone: but it wasn't for lack of effort. There is a moment when a man snaps and all of the rules and regulations and cameras in the world can't control the bloodlust that possesses him.

Coach's pep talks have worked. We kick their ass at practice, then we kick their ass in the game. Revenge is sweet. Coach wears a satisfied look and lets us know how proud he is that we stood our ground and out-toughed the big, bad Cowboys. The locker room feels like we've won a regular season game. Pride is a powerful thing, even when the game doesn't count.

The next week we play the Packers at home in the third preseason game. The third preseason game closely resembles a regular season game. The starters play into the second half and Coach wants to see us clicking. In the fourth and final preseason game, like every year, the starters will rest and the bubble guys will battle it out for the few remaining roster spots. That is usually me: ass clenched and paranoid. But this year is different. I'm not a starter but I'm a contributor. I'm on the field as our first offense battles the Packers' first defense. We are moving the ball with ease. Coach calls a three-tight-end set down on their five-yard line going in. The play is called "spacing." It evenly spaces out their predictable redzone defense with strategically placed receivers. They can't cover all of us.

We are in "Bunch Left": three receivers tight together three yards off the hip of the left tackle. Jay calls for the snap and I run my hook route and sit down on the goal line. He fires a bullet between two defenders and it sticks to my gloves. I fall horizontally on the chalk line for a touchdown. I jump up and, holding the football like a discus, turn and fling it into the south stands. The ref ducks. I have always wanted to do that. Someone up there catches it, I presume. I don't see it land. I high-five my teammates and head to the sidelines to get in the huddle for kickoff coverage.

The next week we travel to Arizona for our last preseason game. Pat pulls me aside before meetings and tells me that I don't have to play in the game the next night. I'm safe.

Coach Shanahan extends the curfew of the guys who aren't playing so we can blow off some steam. We have until 12:30 a.m. Dave Muir, who moved to Arizona after retiring from coaching, picks us up from the hotel after meetings and we head to a local gentleman's club. There's a wild monsoon that night. Thick sheets of water fall down from the sky like gunshots, flooding the streets. But we press on. Our mission is true.

We park in a puddle and run through the deluge to the campdick magnet. Once inside we run into some teammates seated near one of the stages. This particular club hands out popcorn to its patrons, and one of our friends, in a fit of adolescent sexually frustrated excitement, is throwing the popcorn in the air and jumping up and down. We sit on the other side of the room.

I lean back with my beer and take it all in. Blacklit G-strings shoot through my retina and coat my battered brain. The well-rehearsed gyrations of the dancers soothe my aching muscles. I munch a handful of fresh popcorn and finish my Bud Light. Another round. The slightly less naked waitress brings more numbing liquids. Cheers, boys. We clink bottles. The echo of the faulty sentiment summons a smart lightning bolt that tickles a local power line and cuts the power to the club with an audible downshifting "click": pitch black and complete silence. No lights. Naked women. Crazed men. A cacophony of zoo-like snorts and chortles cuts through the silence as the men in the crowd assess the moment. Recognizing the danger, the bouncers yell for the girls to return to the dressing room, which is followed by groans from the men, who had hoped the power outage would break down the walls of strip club etiquette and the dancers would finally be free to ravage the bodies of the patrons they were quietly lusting after, under cover of darkness.

On our way out I make friends with a dancer. I stand near the entrance smiling as she rubs herself on me, inspecting me, telling me with her polished stripper wit about all of the reasons why life is good and nothing ever changes. Professional athletes are attracted to strip clubs. This is well documented. But it's not because athletes are rich, horny animals who gain pleasure in objectifying women. It's because both strippers and professional athletes live on the fringes of a society that judges them for their profession, based solely on stereotypes. These stereotypes are nearly impenetrable. Both stripper and athlete stand alone behind them, and often find solace with those who know what it's like to be there.

Also men like boobs. Football players are included in this demographic. And our hasty retreat to an Arizona strip club, despite the best efforts of Mother Nature to prevent it, is a product of the pendulum swinging back in the other direction. Campdick has me by the balls. I just need to smell a woman, even if she is wearing nine-inch heels.

A few days later Coach Shanahan throws a party at his home to usher in the regular season. Everyone is in good spirits. It represents an important landmark. This is our team: we go together henceforth. This party is our chance to let our guard down and appreciate the moment. Our girlfriends and wives are with us, buffering out the macho posturing. We can relax and mingle with our coaches in a less stressful setting. As the party winds down, I find myself talking with Coach Shanahan near the front door. He tells me that he is really excited about this team. He loves all of the guys. He says I'm doing a good job out there: that he's proud of me.

—Nate, as long as I'm here, and you keep doing what you're doing, you'll be here, too.

Some words are music to the soul. These are a masterpiece; the iron words I've been hoping to hear, or to believe, for my entire career. My hard work is recognized. My skill is respected: by the man who controls my fate.

Just then Sara walks up with her coat and we say goodbye. It is hard to get to know your NFL coach as a man. The environment prevents such intimacy. But slowly, over the years, I have gotten to know Mike. He is a good man. On the ride home I feel myself relax, if only for the night.

Our first game of the season is a Monday night game in Oakland. Our pilot sizes up the island runway at San Francisco International Airport. I look out the western window at the clouds rolling over the green hills of the peninsula, richly lit by the setting sun. As the airplane approaches the runway at SFO, it looks like it will land on the water. It is a majestic scene: the bay directly underneath and the hills and the sky in the distance. The wheels hit the ground and a surge of nostalgia shoots up through my seat. I pause as I reach the open door and look out on the tarmac's fleet of buses: the smell of home fills my lungs. When I reach the bottom of the stairs I stop again. Someone bumps into me. These hills are my hills. Menlo College is fifteen minutes away. So many memories, so many emotions, dreams, pains, and ecstasies: painted in the blood-orange light crawling over the hills. I turn to Tony as we stride toward the bus.

—Isn't it beautiful?

—What?

—What? This! This place! Isn't it beautiful?

—Hell no. This place sucks.

No matter what city we are playing in, there is someone on the team experiencing a similar emotional homecoming and feeling similarly unable to grasp the moment. There are human subplots to every football game—ethereal manifestations of childhood dreams—but we rarely notice them because there just isn't time. There is a football game to play. Then there is another. Always another: next week, next play, next season, the next opportunity for glory. And with this we are stuck in a purgatory of sorts: too far ahead of the journey we can't appreciate, too far behind the glory we can't catch.

The next afternoon I get on the number 3 bus with a stone face. Fuck nostalgia: it's game day. We drive over the San Mateo Bridge from our hotel near SFO and up 880, slowly pulling up to the outskirts of the Raiders tailgate area. A woman on the other side of a chain-link fence separating the parking lot from the road sees us and sneers. She reaches into a cooler and removes a handful of eggs, cocks back, and lets them fly. The first egg misses. The second egg cracks against the window of the seat directly in front of me, where my teammate sits absorbed in his music. The embryonic explosion six inches from his face sends him careening into the aisle of the bus. He recoils in horror as a glob of never-gonna-cluck oozes down the glass.

We make a right and pull through devil's alley: the long, narrow road that leads to the bowels of the stadium. The zombies swarm us like, well, zombies. The music pulses through my headphones: background noise to a delightfully rowdy scene. The line for the port-o-potties usually indicates the drunkenness of the fans. In Oakland the lines are the longest: a piss-taught collection of venomous derelicts snaking through the asphalt. A night game means five extra hours of drinking. The parking lot is a pit of vipers.

Because of the scene in Oakland, my mother stays at home and watches the game on television, even though it's less than an hour's drive from our house. The Black Hole is no place for the mother of the enemy. Raider Mamas are treated like queens. Bronco Mamas: *prima nocte*. But I still had to get more than thirty tickets for friends and family. People assume that we get free tickets to the games but we don't. We get two complimentary tickets for home games; the rest we pay for. For away games we pay for all of them. Taking hits on game tickets is part of the gig. I get ticket requests for every game I play in, and it's rarely accompanied with an offer for reimbursement. People assume. People always assume. The football player obliges and goes broke.

We script the first fifteen plays of every game, a staple of Bill Walsh's west-coast offense. We get the script the day before the game so we know what is coming. One of the early plays is a tricky little comeback route to me on the sideline. It's about a fifteen-yard route. During warm-ups I play it out in my head. Oakland's field doubles as a baseball field. The dirt behind second base comes onto the field near the Raiders' sideline, right at my break point on the comeback route, assuming we are headed south. I really hope we aren't headed south.

But we are. On our first possession, I run my route and make my cut in the dirt. I'm wide open. Jay throws a perfect ball. I feel my feet chopping through the pebbles and dirt clods, trying to stay inbounds. I glance down at my feet just as the ball hits me in the hands and bounces to the ground. Most things I ever did on a football field had to be forcefully learned through hours of painful repetition. Running routes, blocking, tackling, reading defenses, playing special teams: none of these came naturally to me. But catching the ball was easy. Catching the ball was like breathing. I never had to think about it. On the rare occasions that I dropped one, I was mind-numbingly embarrassed.

Compounding my shame were my pleas the night before for all thirty-plus of my slacker friends and family to be on time because I was catching the ball on one of the first plays. Jay glares at me as I run back to the huddle. It brings up third down. Brandon Stokley scoops up a low throw and keeps the chains moving. He makes money catch after money catch, game in, game out. I thank him profusely. We go on to rout them. Late in the game I catch a flat route near their goal line. As I turn upfield their safety corrals me and folds me in half sideways, like a paper clip. I feel my organs squeeze between my ribs and touch on the other side, then snap back into place as I crash to the ground. I jump up and pretend I'm fine.

As the clock ticks down, I stand on the sidelines and watch the circus in the stands. The Raider fans are by now a drunken, blithering mess. A few fights break out and someone catches a shank: nothing a little superglue won't fix. I'm able to observe it all from the safety of the field, separated by metal barricades and the walls of sobriety. I've grown to appreciate the Raider fans over the years. Sure, they're a little rough. But the roughness is authentic and comes from a place of hope. The roughness is born of a common identity with the Raider image, the spirit of rebellion and defiance, toughness and edge, and felonious blade-wielding.

The next week we go on to beat the San Diego Chargers at home. They are now our bitter rivals, thanks in part to a media-generated feud between our quarterbacks, Jay Cutler and Philip Rivers. We are down by seven with under a minute remaining. We get the ball down to their one-yard line. Jay drops back to throw and the ball slips out of his hand. It's a fumble but the whistle blows the play dead. Once the whistle blows, the play is over, even if the video evidence contradicts the call. We get another chance at it and Jay throws a touchdown to Eddie Royal. Coach Shanahan senses the momentum and we go for two: a ballsy move, and one that most coaches are afraid to make. We run the same play with the same result and win by one point, putting our record at 2-0. After the game the replay of the "fumble" runs in a slow-motion loop on ESPN as talking heads in clown suits call for referee Ed Hochuli's job. The advent of replay technology has turned the industry into a bunch of bitching know-it-alls. But football does not happen in slow motion replays. It happens in real time: on the razor's edge. Ed lives on that razor blade. Studio prophets live on a butter knife.

My shoulder separation is an afterthought by now. Now I'm nursing a pectoral strain. It's not a big deal, and gives me a "legitimate" excuse for getting an injection the night before the game. This pacifies the subconscious reluctance of both doctor and patient to engage in an overtly risky and unsound medical practice just to juice up a bit player for an early season football game.

Boublik:

> The player is asking about a Toradol injection in anticipation
> of tomorrow's game. He states he has some residual soreness
> in his right chest wall. After discussion of risks and benefits
> of Toradol including risk of infection at the injection site,
> bleeding, liver damage and kidney damage, the player is given
> an injection of 60mg of Toradol IM into his right buttock
> under sterile technique. This is well tolerated.

The next day we play the Saints at home. I go up high in the first quarter and catch a one-yard touchdown in the north end zone. Tony and I do the jumping butt-touch celebration. We win the game and move to 3-0. One week later during Friday practice, I run a hook route and feel something yank in my right oblique area. It's toward the end of a light practice. It is diagnosed as a "right costo-chondral irritation at roughly the 10th rib." Make sense?

The next morning we leave for Kansas City. I'm in a lot of pain, possibly the most painful injury I've ever played with. Another 60 milligrams of Toradol into my ass the night before the game but it doesn't help. The muscles in the torso are constantly at work while playing football. Twisting, cutting, exploding, sprinting: all of it activates the obliques. Warm-ups are so painful that I'm considering the unthinkable: telling coach I can't play. The Toradol, the adrenaline, and my access to the pain switch: none of them can override this invisible injury. But my pride won't let me pull the plug. I suit up and tell myself, once again, that I am a warrior, and this is my war. I stare at myself in the mirror and fight back the fear. It is dangerous to be on an NFL field if you're not healthy. Trained killers are coming for you. As I run on the field before each play, I ask myself: how are you going to get through this? And after each play, I ask myself: how are you going to get through the next one? Eventually the game is over.

More Toradol for the next week's game, all subsequent games, and all previous games. Every game a needle.

• • •

On a Thursday night game in Cleveland a month later, I'm in the slot on the left side of the ball. The fourth quarter is winding down. Jay snaps the ball and I take off up the seam, bending in toward the middle of the field. I see Jay cock back and throw the ball in my direction. Now it is mine. I must catch it. Catch the brown rainbow. Millions of people are watching, but they don't exist. I'm alone again inside the timeless moment of football chaos. I give one last grunting burst and leave my feet, shooting out over troubled waters. The ball sinks into my fingertips. I curl my fingers in toward my palms and—*CRACK!* An M-80 explodes in my helmet. The hit knocks me out for a moment. I get off the field and we win the game.

I'm dizzy and depressed and my neck is locked for the next week. But I don't receive any treatment for it. By then I know the drill. Come in to work and get strapped to machines all day so we can log it in the book. I weigh the options in my head: peace of mind or peace of nothing. I choose peace of mind and stay at home where I can rest and medicate myself with drugs of my choosing, drugs that don't come out of a needle and won't eat away my stomach lining.

Another routine 60 milligrams of Toradol for the following game in Atlanta.

As the season wears on, my weight drops further and further below acceptable levels for my job description. It's hard for me to keep on the tight end pounds, because practices are so strenuous, my metabolism is so fast, and I'm never hungry. The weight loss compromises my ability to block defensive linemen. I'm getting thrown around, so every once in a while I take a scoop of creatine to help build muscle and put on weight. But the creatine dries me out. I need to drink a lot of water; otherwise I'll cramp and be more susceptible to muscle pulls. But I prefer that to getting my ass kicked every day.

The next week, during practice, I accelerate to track down one of Jay's balls and my right hamstring, the one that had the "mild hamstring strain" three years earlier, fails me once and for all. Season over again. Career soon to follow.

The Lighthouse Diagnostic Imaging Center reports its findings:

There is a complete tear of the contouring biceps femoris semitendinosus from its origin at the ischial tuberosity. The fluid-filled defect is appreciated well on axial images 11. The semimembranosus component is intact. The torn tendon is retracted slightly greater than 2cm as next seen on axial images 14. There is a mild amount of surrounding peritendinous edema. There is intramuscular and peri fascial edema primarily involving the semitendinosus muscle. The more distal biceps femoris component of the tendon is thick and low signal. This suggests chronic tendinosis and prior injury.

Comparison with large field of view images of the pelvis from November 2007 confirm chronic right-sided hamstring tendinosis with tendon degeneration or partial tearing affecting the biceps femoris semitendinosis on the prior examination.

There is chronic partial tearing involving the adductor longus muscles from the pubic symphesis [sic] bilaterally, with greater involvement on the left. The distal fibers of the rectus abdominous [sic] are intact. There is fluid in the symphysis pubis with secondary cleft extending along the right aspect of the symphesis [sic]. This does not appear to be an acute injury.

Not an acute injury. Both groins already torn. Both hamstrings already torn. Both hips already torn. The hamstring that had bothered me for years was torn from its attachment the whole time, and no one ever told me. It had never healed from the rehab or the injection or the off-season or anything. It was ready to blow.

I go back to Vail for another PRP injection and lie on my face as one more needle pushes down into my soul. I start the maddening rehab process once again, lying around the training room and trying to figure out why it happened. Was it the Toradol shots? All the anti-inflammatories and painkillers? My diet? Was it the creatine? Poor treatments of my chronic hamstring injury? Poor health care in general? The steroid injection in the ischial tuberosity three years earlier? Was it the hamstring overcompensating for a weak groin? My weight gain? A weak core? Fatigue? Was it my mind? Fate? God? No. None of it. It is football. I play football for a living.

I stay on my couch that week and watch us lose to the Raiders at home. Never lose to the Raiders: especially at home. It's an ominous sign. We win the next two games and get to 8-5: in total control of our playoff destiny. It's hard to win in the NFL. If you make the playoffs, you're a great football team. Denver has come to expect that greatness because Coach Shanahan has consistently delivered it. But we have missed the playoffs in the previous two seasons. The natives are getting restless. All we have to do, though, is win one of our last three games and we are in the playoffs. Reach the playoffs and everyone should be safe for another year in paradise. But fate has other things in mind.

I lie supine on a back corner training room table and watch a tide of poison molasses roll over our team. I can feel the life being sucked out of us. I stay at home while the team travels to Carolina and gets rolled. That sets up a home game against the lowly Bills in week fifteen. It's all too easy: too perfect. Win and we're in. I stand bundled up on the sidelines in the frigid Colorado winter and watch our season swirl down the drain. We are powerless to stop it. We lose 30–23 to a chorus of lecherous boos, setting up a week-sixteen matchup in San Diego against the surging Chargers. I go along on that trip to support the guys. But we don't even need to show up for that one. It's preordained. We get steamrolled. Our season is over.

A few days later Mike Shanahan is fired as head coach of the Denver Broncos. Not long after that, I am fired as backup tight end/special teams player for the Denver Broncos. Look, Ma, I'm nothing.

12

The After Affect
(2009)

A week after being cut, I fly back to Denver to clean out my locker and say goodbye to my friends who work for the team. All of my teammates are gone for the off-season. I'll never see them again. Flip and the guys in the equipment room, Greek and Corey and Trae, and Rich and Crime, and everyone else. They have become my extended family. When I came to Denver, I came alone. All players do in one way or another. The Bronco organization was my lifeline. They were very good to me. I love them. I want to tell them how I feel about my time there. But I don't have the words.

All I can think about is Josh McDaniels not calling me back. I want to run into him in the parking lot. I won't need any words for that. I have a bone to break with him. But Flip tells me that he's not even here. He's in Indianapolis for the combine. Lucky Josh. Not that I don't understand his indifference. He's thirty-two years old. He's just suicide-squeezed his way into the head coaching position of one of the NFL's most venerable institutions, taking over for a future Hall of Fame coach who controlled the entire operation from top to bottom. The last thing he wants to do is waste his time explaining to a backup tight end why he doesn't fit into the plan.

I sit down in my locker for the last time. It was always a bit out of sorts, full of clothes and shoes and tape and gloves, notebooks and letters and gifts. Do I even want these cleats? These gloves? These memories? Yes. I fill up my box. Six years as a Denver Bronco. Six more than most people can say. Still feels like a failure, though. So this is how the end feels? Standing in an empty locker room with a box in my hand? Yep. Now leave.

I get home and call Ryan. He knows that my prospects aren't great. I am an undersized tight end with injury problems and I am pushing thirty. I need to find a team that wants a player with my skill set and won't be turned off by the injuries. That won't be easy, especially because the most recent one hasn't healed. What I couldn't convey honestly to Greek I can to Ryan. There is a problem—a deeper problem—that's affecting my body. It's not simply that my hamstring is shit. The entire functional movement of my body is off. I can feel it with every step I take. Something is amiss.

Ryan sets me up with a biomechanics specialist/physical therapist in San Diego named Derek Samuel. Ryan thinks I'll get along with him. He'll assess my situation and we'll go from there. But I'm afraid this won't be enough. Desperate times, you know the saying. I reach out to a connection I made a year earlier and acquire a supply of human growth hormone, HGH. The drugs come in the mail in a package stuffed with dry ice. I half expect to see the feds storm out of the bushes, guns blazing, as I pull the box off my front porch.

But no feds. Just me and another needle.

It comes with very little guidance as to the quantity and regularity of the shot. I have a conversation with my supplier and he tells me how to do it. Other than that I'm on my own. I will tell no one what I'm doing. I go to the store and buy syringes and start injecting it in my stomach immediately. I am paranoid about every aspect of this decision. I've never used performance-enhancing drugs. Haven't ever even seen them. I take pride in my natural ability and I don't want to taint it. I don't want to test the karmic winds. But

I also don't want to taste the death of my football dreams, not like this.

I pack up my Denali and head over the Rocky Mountains, the vials of HGH stuffed in an ice-filled cooler. My time with the Broncos is up. That's for sure. The rest of it will reveal itself eventually. But all men must move along. And they must do it with the feeling that they have left business unfinished, relationships unformed, opportunities untaken. I played for the Denver Broncos. I achieved my dream, which confronted me with a naked truth: the dream has been won, and it is not enough. I leave for San Diego to revive the dream, to give it the fresh air it needs, so that I can leave the game on my own terms.

From the moment I step into Derek's La Jolla office, which connects to a small fitness club, I know I am in the right place. Derek is a six-foot-three former volleyball player at the University of California, Irvine, with a friendly disposition and a freaky knowledge of the human body. He asks me questions and lets me talk. He is interested in what I think and not just what I was told. He is interested in how the treatments I got in Denver affected me, how I responded to them and how I felt about them. He wants to know the backstory so he can make more sense of what he discovers on his own. It is refreshing, truly. For the first time in years I am free to look at my body through my own eyes and to own my own flesh.

After an examination, Derek determines that not only is my hamstring incredibly weak, but my hips are drastically misaligned, my pelvis is tilted forward, and my core strength is very poor. We get to work immediately, realigning my body and strengthening its foundation: the core. This isn't accomplished by snapping it back in place. It takes a soft touch, a gradual redirection of the years of bad habits that I formed while playing football. The body must correct itself. And it must be listened to every day. Derek's genius lies in his ability to hear the human body's cries. But in order to do that, he also has to hear the cries of the mind. He is simultaneously acting as physical therapist and psychotherapist, easing me into a transition I am denying. I think my football days are far from over, yet I rail every day about the oppressive nature of the industry. Between sets of exercises, I go on and on about the meat market of the NFL, the hypocrisy of coaches, the false glamour of fame, and the inevitable meltdown of the players who play football.

He sees many football players in his practice, and knows firsthand what I'm talking about, but he never lays it on me with any air of finality. He is too smart for that. But the fact that I am here at all, seeing Derek for my treatment, means that I have cleared an existential boundary in my pursuit of football absolutism. I have turned my back on the modern philosophy of NFL injury treatment, and in doing so, have taken one more step away from the industry I think I am running toward. Certainly, I am getting superb treatment. But my mind is doing something else. It is picking apart a system to which I have bowed my entire adult life. It is, as a defense mechanism I suppose, finding all of the reasons why I should cash in my chips and walk away. But I can't.

Along with Derek's workouts in La Jolla, I am meeting a track coach at the University of California, San Diego. My first day on the track is a sad realization of how unhealed my hamstring is. To be medically cleared by the Broncos, I had to pass a series of strength and endurance tests. But those tests did not include running. I haven't run since the corner route that tore it three months earlier. And when I try to open up and sprint on the track, I can't. I simply cannot run. Judging by the look on the track coach's face, it's a sad sight. I wish the drugs would fix me already.

I've been hiding the vials in the refrigerator of my friend Billy, who I'm staying with in San Diego. I become a master of refrigerator organization with the express purpose of concealing contraband, wrapped in the folds of my deli meat. Nobody touch my *goddamned turkey*!

When it's shooting time, I get the turkey and retreat to my room, where I mix the active ingredient with the solution, pinch my belly, plunge the needle into my skin, and push the poison into my body.

Once again, I imagine the HGH as a fleet of noble warriors. I am doing God's work, after all. There is nothing dishonest about it. HGH is a hormone naturally produced in the body, and my body is starved for it. I am broken. I am unable to perform the sacred task that I was born for. To turn my back on it would be to spit in the face of God. So I draw back the syringe and poison myself again, and again, and again. I don't know what I expect to happen: a miraculous recovery perhaps. A newfound clarity. A bigger dick. But nothing happens. The only noticeable effect is that my body aches viciously. I don't know what this means and I'm too afraid to ask anyone about it, so I assume it is part of the process. My muscles are big and well defined, perhaps slightly more than normal, but perhaps not. I was always muscular, and I am training like a madman. My body looked the same when I was in Denver shooting tequila instead of HGH. Either I'm not doing it right or my body doesn't like it. And the fact that I'm doing it at all, sneaking around, carrying this secret, makes me feel mentally weak and undeserving of good fortune. Regardless, after a few months of intrastate trafficking and refrigerator espionage, I throw it all away in one dramatic garbage-can dump, and I close the lid. If I am going to play football again, I'll do it clean.

About that time, in April or May 2009, I start receiving calls from Eric Van Heusen, the tight end coach with the Las Vegas Locomotives, one of six teams in the upstart United Football League. The UFL is hoping to steal some of the NFL's thunder. Alternate leagues have been tried in the past and all of them have failed. The branding of the NFL is too good. But the financial backers behind the UFL have an idea that the NFL might implode over the collective bargaining labor dispute and leave the country scrambling to fill the void the NFL's absence created. It is a worthy gamble, I suppose, if you have money to burn.

The Locos have drafted me and now hold my rights. Jim Fassel, the former head coach of the New York Giants, is to be the Locos' head coach and they have already received commitments from a host of former NFL players in limbo, like me. I tell Van Heusen that we'll talk about all of that if the NFL doesn't work out. But I believe it will work out. And when the call comes I'll be ready.

So I keep training. My body stops aching, and thanks to Derek's training methods, I am getting very strong. I feel pretty good considering the hamstring. And I'm getting my stride back on the track. My days consist of four or five hours of training. That's all I do. Ryan tells me to stay ready. Teams have shown some interest and once training camp starts, men will be falling and they'll need a veteran to come in right away and be ready to play. Well then, I'll be that veteran. I'll be ready!

I run sprints on a field that overlooks La Jolla cove; I'm sweating in the San Diego sun and trying to manifest my destiny. When I sprint, I feel vulnerable. I feel like I will snap and collapse into a bag of bones at any moment. But I can sprint. Not as fast as I once could, but fast enough for a tight end. And perhaps the hamstring isn't so bad after all. Perhaps it's healing.

The summer creeps by and I try to stay positive. I convince myself that things will work out if I keep working hard and trust my instinct. Never mind that my instinct, my true instinct, has been screaming at me for years. But I have stopped listening.

One day after workouts I go into an American Apparel store in San Diego and there is a smoke detector beeping. It has a battery that needs to be changed. It's a loud, sustained beep every fifteen seconds. A month later I return to the same store. And still the same beeping. Only now it's two of them. And the same girl is working behind the counter. I buy two T-shirts and as I am paying for them, I ask her about the beeping.

—What beeping?

What beeping? She can't hear it. She's just like me. My body beeps all day long.

Training camps start in late July and I'm not on a team. But I have no other plan. I will train and I will wait. Someone will call. Meanwhile, the therapy sessions with Derek are unearthing more misgivings that I have with the NFL. I rail against what I now see as years of mishandled injuries, against the emptiness of fornicating with the jersey chasers, against my own inability to turn from the game, against my monetary motivations for still wanting to play it, against the media's petty ownership of the players, and against the entire bastardized commercialization of what to me is the most beautiful game on earth. And here is the crux of it: I still believe in the beauty of the game. This above all else is true. But to be a fly on the wall, or to be Derek, is to be struck in the face with how delusional a man scorned by his lover can be. Here I am telling him all the reasons why I hate her, in between sets of an exercise specifically designed to lead me back into her arms. I am sick.

Then, in early August, my phone rings. It's the Philadelphia Eagles. They want to fly me in for a workout. Later that day I am at the San Diego airport on my way to Bethlehem, Pennsylvania, where their training camp is being held. I am picked up at the airport by an Eagles employee, a college kid with an internship. I hop into his college kid car, scattered with papers and trash. He pulls the pile of stuff from the passenger seat and tosses it onto the pile of stuff in the backseat. I kick some water bottles aside to make room for my feet and we drive through the city toward my hotel. There is a music festival in Bethlehem that forces a long detour, which gives us plenty of time to think of stuff to talk about. Girls. Yes, girls. We both like girls.

The next morning I eat a light breakfast and go to the lobby at the scheduled pickup time. Also milling around the lobby are two other dudes just like me. This is how it will be. A team needs a player, so they fly in several of the same position and work them out, and they take the best one. There are three of us for this workout, and while we get ready in the locker room, we make small talk.

—This is pretty weird, huh?

—Shit, this is my third workout in the last week.

—It's my second.

—Damn. This is my first.

—But you played before, right?

I was obviously the elder statesman.

—Yeah, for the Broncos.

—Dass whassup.

—You guys?

—Naw, I just got cut last week.

—Naw, me neither. I was in minicamp in Cleveland. They cut me the day before training camp started.

—What? That's dirty.

—Who you tellin'? I seen it comin', too. They was treatin' me like a damn stranger.

—Damn. . . . You're both rookies?

—Yup.

—Yeah.

Eventually we make our way onto an adjacent practice field, where the Eagles are finishing up morning practice. After we warm up, a scout calls us over and runs through what the workout will consist of: a few blocking drills, some receiving drills, then we'll run routes and catch balls. The workout goes as smoothly as possible. My adrenaline is racing and I feel good. I run crisp routes and I catch every ball that's thrown to me. But instead of having an actual quarterback throw us passes, they found a quality control coach who convinced someone that he had a good arm. He didn't. He was accurate but he had a noodle arm. There was no velocity behind his passes so there was no opportunity to show off my ball skills. Every ball floated down into my hands like a feather. It was the same with the other two guys. We all looked good. But I am certain I'll be signed on the spot. They are both rookies and I am a six-year veteran. Compared to them, I have no learning curve, and I can come in right away and contribute on special teams.

After waiting around outside the trainer's offices, we go through a physical with their medical staff. They ask me about my past injuries. I tell them that everything is 100 percent. No problems. They take their notes and send me on my way. Then just as quickly, we pile in the van and leave. We are almost off the college campus when our handler receives a call.

—Yep, yep. Yeah, he's right here. Okay. Okay, we're coming back.

He turns the van around. They want me to come sign my contract. My heart speeds up in anticipation of a new beginning in Philadelphia. What will my life be like in this new place? How will I fit in with my teammates? Will my body hold up? We pull back into the parking lot and our driver turns to us in the back. I sit forward in my seat.

—Rob, they want to talk to you inside. You can take your bag.

Rob nods to us, gets his stuff, and walks toward his new job. We drive away. I lean back. An hour later I'm at the airport, rushing to catch a flight they booked for me thirty minutes earlier. I squeeze into my middle seat, put on my headphones, and shake my head. What the fuck am I doing?

The Cleveland Browns call me three days later. Same story. They want me on a flight that night for a workout the next morning. But it's already evening when they call. So they put me on a red-eye. I'm picked up and taken directly to the facility, where I change with one other guy in the coaches' locker room and we make our way to the indoor field for our workout. Also at the indoor field is half the Browns team, going through a stretching routine intended to flush the soreness and the gunk out of the beaten bodies of training camp. They are doing their stretches, but they are watching us. I know it's likely that somewhere among those nameless faces, some tight end sees us and knows that he is in trouble.

The workout is similar to the one in Philly, except for one addition: the forty-yard dash. I am prepared for it but quietly hoping like hell I don't have to do it. I haven't run the forty for time since Ryan and I met the 49ers scouts at Stanford the day before the draft seven years ago. It takes a good amount of technical training to be a good forty runner. And it's not football, either. Football is never a straight line out of a track start. And my hamstring is shit. I can mask its shittiness as long as I don't have to hit top-end speed, which is actually easy to avoid on a football field. Football players rarely hit fifth gear. But the forty-yard dash requires it. And I'm worried my entire pelvis will explode right there on that field, with scouts timing me and coaches evaluating me and players watching me. Kaput. Get him a body bag.

But my pelvis doesn't explode. I run a 4.6: plenty fast for my job description. The rest of the workout goes great. I catch everything. I even make a few improbable circus catches that I know no one else can make. After the workout, they bring me up to a coach's office area and ask me to wait for a while until I can do a physical and meet with the GM. They send the other guy home. I sit in the room for four hours watching daytime television. Finally I am brought downstairs for the physical. They poke and prod. I tell them I feel marvelous. Then I go back upstairs and meet with General Manager George Kokinis in his office. It overlooks the practice fields. He is wearing standard-issue Browns coaches gear: visor, team T-shirt tucked into mesh shorts, ankle socks, and tennis shoes, sitting at a huge mahogany desk, pictures of his family everywhere, a large television with practice film paused and a remote control on his desk.

—Well, Nate. I can tell you're a good player. You can play in this league. We just have to find a spot for you. You play special teams, right?

—Yeah.

—Do you remember what games were your best games? So we can watch the tape and I can show it to our special teams coach.

—Umm. Let me think. Uh, maybe the San Diego game. Uh . . .

—Yeah, you think about that one and get back to me.

—Okay, I will.

—So we're going to send you home today, but be ready. We could bring you back any day. It could be tomorrow, next week, whenever. If we can make it happen, we will. Make sense?

—Yeah, makes sense.

—Great.

We shake hands and I am back in a van. Then at the airport. Then squeezing back into my middle seat and repeating my new mantra: What the fuck am I doing?

This is a side of the NFL I am not used to. I knew of it, certainly, but I didn't know what it felt like. And it feels damn awful. But five days later I get a call from George Kokinis. He found me a spot. Well I'll be a monkey's uncle. I hop on another red-eye flight and am picked up from the airport in the same van driven by the same dude from last week: déjà vroom, straight to the facility. I sign my contract, eat breakfast, get my gear, am issued my locker, and before I know it, I am wearing my number 85 Cleveland Browns jersey and jogging onto the practice field with my helmet in my hand.

Every team has its different routines. Often the most difficult part of being on a new team is getting adjusted to the way they do things. The team takes on the personality of its head coach, and every coach is different. In this case, it is Eric Mangini. I had heard a good deal about Coach Mangini from a few of my teammates in Denver. We would sit around the table in the cafeteria and talk shop, and several times I heard tales of Mangini's evil. New York, while he was coaching the Jets, was hell. No, not hell. Worse. Three-and-a-half-hour practices. Busted bodies. Jangled nerves. Cussing. Yelling. Tension. Belittling. Football, the game, was nearly unrecognizable under Mangini's demented eye. Hell was no match for it.

But surely the stories were overblown. Their color was more vibrant because of the contrast. They were being told in the peaceful valley of Shanahan: the heaven to Mangini's hell. Mike Shanahan knew how to run a team. That meant he knew how to treat the men on it. Being a head football coach is not about being a strategic genius. Every coach in the NFL knows football strategy. It's about leading a group of grown men toward a tangible goal and treating them with the respect their sacrifice deserves. That's how you get them to play well. Many players, upon arriving in Denver, were flab-bergasted by how well Shanahan treated them.

—You don't know how good we got it here, man.

I always heard it, but I never understood it. Coach Shanahan was all I really knew. He was the model of an NFL coach in my mind. I went through one camp with Steve Mariucci and one with Dennis Erickson, but these were back when I was a boy in the NFL, too consumed with my own performance to pay any attention to the performance of my coaches.

But by the time I arrive in Cleveland, the mystique of the NFL has vanished. My eyes and ears are open. From the blow of the morning practice's first air horn, I know I'm in a strange place. Warm-ups are usually very relaxed. They are designed to get the player's body warmed up, and everywhere I have ever played, the coaches have allowed us to warm up at our own pace, as long as we are ready to practice hard once warm-ups were over. But here in Cleveland, warm-ups are frantic and explosive. There are coach-es barking orders and players are running through bags like Navy SEALs.

—Get your knees up!

—Keep that ball high and tight!

—Come on! Let's go! Let's go!

Oh, brother. This is not good.

As a veteran player gets on in age, he loses his patience for rah-rah rituals that he knows are worthless. Grown men with refined football skills do not need to be goaded and harangued. Football is brutal enough without someone yelling at you. And if you make it to the NFL, you're a self-starter. It isn't high school. You aren't dealing with children. But nobody told that to Mangina.

Practice is long and physical. I spend it standing next to my new tight end coach trying to pick up on the terminology. The Browns offense, led by former Patriots coach/Brady jockstrap carrier, of-fensive coordinator Brian Daboll, is complicated and seems to have no rhyme or reason: arbitrary names for strange concepts. But I have been in the same west-coast offense since Menlo College. I am used to that language. And this system is an entirely different language, so of course it will sound like arbitrary names for strange concepts. But this is the end of training camp. People should know their stuff by now. When I ask my new teammates to explain something to me, though, they just shrug.

—Shit, I don't know what to tell you, Nate.

If they don't know it, I'm in trouble.

I play some scout team offense and do okay. But I'm rusty. I was training hard in San Diego, but I wasn't playing football. My run-blocking technique has fallen to shit. I'm not a natural tight end, so for me to be a good blocker, I have to work on that technique every day. The only way to do that is to practice in pads. As horrible as it is strapping up every day and banging heads, it's the only way for a guy like me to have a chance at blocking three-hundred-pound athletes. I have to knock the rust off quickly.

Meanwhile, I'm catching some weird vibes around the building. Things feel off. I'm focused on learning the system as fast as I can, so I don't have a lot of time for psychoanalysis, but it's hard to miss. To a man, the entire Browns team seems to be deep in despair. There is a natural sluggishness that occurs during training camp, but this is something different. The men seem positively broken. They have no fight left in them. The locker room is quiet, so quiet. In Denver, even in the midst of training camp, the locker room was lively and social. Cleveland is a mausoleum. That night at my first team meeting, I learn why.

As I sit down in the emptiest seat I can find, I notice that players have handwritten notes scattered about their desks and their laps. They are reading them over nervously. Coach Mangini, a doughy thirty-eight-year-old frat boy with parted hair and a butt-chin, walks in and takes his place at the podium, a dip in his lip and a Styrofoam cup in his hand. He starts off by welcoming the two new men who were signed to the team that morning: me and some other dude.

Then:

—To show them how we do things around here, J.P., stand up.

J.P. stands.

—There is a quote written above the door to the locker room; what does it say?

—Uh, you must choose: the pain of discipline or the pain of regret.

—Very good. You can sit down. Clarence, stand up.

—Sheeit.

He says it under his breath. Muffled chuckles from the audience.

—We have six core values on this team; what are they?

—Damn. Okay, um, trust, communication . . . um, hard work . . . umm . . .

Someone whispers from behind him.

—Focus!

—That's right, Clarence, *focus*. Okay, two more.

Silence.

—Come on, *Clarence*. . . . Can anyone help him out?

From somewhere:

—Intelligence.

—Football is important to you.

—Good. Clarence, you gotta know these. And I'm going to keep calling on you until you do. Sit down. B.J., stand up. Tell me the name and number of every offensive lineman on our roster.

B.J. was a rookie defensive back and rattled them off like a pro.

—Okay, good. Very good.

Then Mangini presses play on the video system and footage *of the morning's warm-ups* comes onto the screen. He had the warm-ups filmed and the tape cut up and cued up for the meeting. He launches into a biting critique of each player's warm-up performance, excoriating certain players for not having a sense of urgency during the drills, and referring again and again to the mantras that are written in big block letters around the facility. He preaches the importance of living by their words, and humiliates the most glaring examples of those who aren't.

—You must choose, the pain of discipline or the pain of regret.

—Every battle is won before it is ever fought.

—Don't sacrifice what you want most for what you want now.

And on the training room wall, "Durability is more important than Ability." As if the injured guys don't feel bad enough already. Might as well say, "If you're reading this, you're a pussy." That's what all the notes are. People are making sure they have these fucking mantras memorized. What the fuck is going on here? When the meeting breaks, I track down a fellow tight end.

—Is he *serious*?

—Yes, dude. Dead serious.

Aside from the food, which is delicious, Cleveland is hell. Practices are long and tense and confusing. Meetings are confusing. There are "voluntary" meeting sessions for rookies and new guys, called "Football School," which are also confusing. The players are depressed, myself included. Also my body feels awful. The first few days of practice were okay but by the third day I feel like I'll snap at any moment. My knee is bothering me for no reason. My hamstring and hips are tight. And to top it off, I have absolutely no idea what is going on in the offense. It is a completely foreign language. And no one is teaching it to me. My only chance is to get in good with the special teams coach, and he can't be bothered. It's strange that I'm even here.

Luckily, there's a game to prepare for that breaks the spell of practice hell. I had arrived on Monday morning, practiced all week, and by Friday night, am hoping maybe I'll get in the game the following evening. I don't know much, but I know enough to get by. And the quarterbacks are nice guys. They'll help me if I need it.

The night before the game, I check in to the hotel and go down to the meal room. Again, the food is amazing. I am blown away by it. There are artisan chefs stationed around the room creating made-to-order delicacies: everything you can imagine. Pastas, Mexican food, omelets, salads, a variety of roasts, meats, grains, fruits, breads, cookies and pies. It makes Denver's food selection look like the HealthSouth cafeteria. After dinner is our team meeting. And here comes Mangini again, same smarmy look on his face, same paranoia in the crowd. Only now I'm among them. I have notes scattered around my lap, too. My heart is racing. *Please don't call on me please don't call on me.* He calls on a few guys and has them stand and answer more arbitrary questions about the Titans' defensive tendencies and historical success running certain coverages and substitution packages and, holy shit! It's embarrassing. I breathe a sigh of relief when he concludes the question-and-answer portion of the show and moves on.

Then he motions to a young man in army fatigues standing in the corner of the room and introduces him as an Iraqi war veteran. Coach wants him to say a few words to us. The football-as-war metaphor is an old motivational tactic. I have heard it evoked many times in my life. But not like this.

The vet tells us his story. He lost three friends and both of his legs in a roadside bomb attack the previous year. You can hear a pin drop. He's an impressive man, an impressive kid, really. But like me, he seems confused as to why he is here, addressing a room full of professional football players the night before a preseason game. It soon becomes apparent why he was brought here. Mangini starts peppering him with leading questions intended to strengthen the validity of his own mantras, trying to draw an honest parallel between the bomb that killed his friends and the following evening's preseason game against the Tennessee Titans. The soldier sees what Mangini is doing and steers away from it, choosing instead to speak candidly about what he had learned, not what Mangini had hoped he learned.

After a few cringeworthy questions from the audience, class is dismissed. I make a beeline to my room, where I lock myself behind the double bolt and scribble furiously in my notebook. This is some outlandish shit. And I don't want to forget it.

The next night we play the Titans. I suit up in my number 85 game-day gear. I look at myself in the mirror before the game, wearing all brown. This color looks strange after years in blue and orange. But I'm in a uniform: I guess that's what matters. The game starts and I am ready but I never set foot on the field. It's just as well. I need another week of practice.

We have the next day off. I go into the facility for a workout, then back to my hotel room. I sit around the rest of the day. Outside is a heavy rain. I stare out the window and repeat my mantra.

The next morning I walk into the facility at around seven. As I open the door, I see the grim reaper leaning on the wall about fifty feet away. The grim reaper is the member of the staff in charge of telling players that the coach or the GM wants to see them upstairs. And bring your playbook. It's the end of the line. The grim reaper was that pear-shaped little penguin-man with the pronounced FUPA on HBO's *Hard Knocks* that the Cincinnati Bengals employed to rouse professional athletes out of their sleep before dawn and tell them they weren't good enough to play anymore. There is an art to being the grim reaper. The penguin was not an artist.

But this grim reaper is. And there he is, leaning on the wall, waiting for his target to walk through the glass double doors. Poor guy, I think. Not the reaper, but whoever he is waiting for. Easy come, easy go, right? As I clear the glass double doors and make my way down the hall, he perks up and pushes himself off the wall. No fucking way.

—Nate. George needs to see you upstairs.

Up the stairs we go to complete the filthy cycle.

I sit down once again in front of that stupid mahogany desk. George hands me a manila envelope with my walking papers in it.

—Well, Nate, I'm sorry about this. We thought you could come in and add a different dimension to the offense. But it's just too close to the start of the season to get a good look at you. I have no doubt you're a good player, but you'd be better off in a system that . . .

Blah blah blah and on and on he goes. I'm not paying any attention. I am busy bashing his skull against his big, beautiful desk while his family members look on through the foggy lens of forgotten picture frames. But I know it's not George's fault. I like George. He was the only reason I was there in the first place: him and my tight end coach. George went to bat for me and convinced Mangini and Daboll to give me a shot. It was those two who decided I was no good. George just had to be the one to tell me. Yes, this is all part of the business. Yes, it's what I signed up for. I should be happy that I got to be a part of it at all. Look at this! I was a Cleveland Brown! That's more than most people can say. I am a lucky man. I should be thankful.

But thankful for what? Thankful that I was given the talent to play the game I love? Yes, I'll buy that. Thankful to be subjected to the whims of the men who control the game I love? Hardly. There are thousands of George Kokinises and Eric Manginis in the football world, men who love the game but weren't good enough to play it, so they found a way to control those who are. They are trying their best to build a perfect football team, yet they're losing the perspective needed to do it. And they're polluting the stream that every football-loving child in America is drinking from. They've forgotten about the players. A coach is only as good as his team *feels*. And if he doesn't have their respect, what does any of it matter?

I go back home to Denver this time and swan dive into the pit of despair. A week later, the day after the last preseason game of the summer, I get a call from the Saints. Their second tight end, Billy Miller, tore his Achilles tendon in the game and they need a veteran to come in right away. They fly in four of us, all of whom have at least five years of experience. I know two of them well: Jeb Putzier from Denver and Daniel Wilcox from NFL Europe. We are all veterans, we are all in good shape, and we all want the job.

The workouts for the Eagles and the Browns were led by an assistant coach who also served as the quarterback. But in New Orleans, head coach Sean Payton is leading the workout and quarterback Drew Brees is throwing us the passes. It feels more important. I have another solid workout. I run another 4.6 and catch everything. But we all do. And after we shower and change, we all hop back in the van and are dropped off at the airport. None of us gets the gig.

The season starts and I'm unemployed. I have to make a choice. UFL training camp is starting in a week. In the conversations I had over the summer with Eric Van Heusen, we talked about a lot of things. One was, well, what's the point? EVH, as I called him, assured me that the NFL would be plucking men off UFL rosters when their own players got hurt because the UFL guys would be in football shape. They would be polished and ready. EVH and the rest of the coaches had to really sell this one because all of the UFL guys believed they should be NFL guys, believed they *would* be NFL guys; all they needed was another chance.

Another thing we talked about was money. EVH said they hadn't worked out the particulars yet but it would definitely be six figures. Well, that is good. Certainly it won't be NFL money, but six figures is good. Playing in the NFL warps your perception of money, and after earning large sums of it to play a violent game a certain way, the prospect of living that same strange, brutal life for a fraction of the reward is unappealing. This was a major obstacle the UFL was facing. Football is fun when you don't know any better. But when your body starts to break, unless you are getting paid well, there is no incentive to continue. There's a reason why you don't see grown men at the park in full pads playing football games. They claim to love the game so much. They claim that the pros are so lucky to play it. Well, you, too, Johnny Crotchscratcher, can play the game you love. Put an ad on Craigslist. Start your own league. Go hit somebody.

When I talked to EVH after getting cut by the Browns, he told me it was going to be more like $80,000. Not what I was expecting but still good for three months of work. After my final workout in New Orleans, whenever I worked out, I broke out in hives. They crawled all the way up my back and sides onto my neck and face.

Large, raised boils and coagulated blood islands formed on the sea of my skin and begged me to reconsider whatever my body knew I was about to do. What beeping? I ignored the cries and pushed on. I was in great shape. I had trained all summer. And for what? To pack it in? Then what would I do with this ax I'd been sharpening? There's nowhere else I can swing it. I call EVH and tell him I'm in.

Great! Oh, but one thing, EVH says, now it's $50,000. That's all they can do. Eh, fuck it. I'll do it. I drive from San Diego to Arizona, where all of the UFL festivities are taking place. First order of business: pick up my playbook. Jim Fassel's Scottsdale home doubles as the Locos' off-season coaching offices and it's where I go to meet EVH and Fassel in person. They are nice guys and tell me they're happy to have me. EVH gives me my playbook and I leave for the hotel in Scottsdale where we'll stay for three nights and go through extensive physicals and meetings and orientations. But hold on! First we have to sign the contract. Get it in ink! We all showed up despite being misled by the numbers. But the big surprise comes in the hotel ballroom, when they hand us the pen. The real salary, the one we'll all be receiving for our services, is $35,000: $35,000 to keep the dream alive. I sign it in blood. Look, Ma, I'm a Las Vegas Loco.

Training camp, and the Locos' home base for the duration of the six-game season, will be located in Casa Grande, Arizona. One hour southeast of Phoenix. Population: 48,571. Climate: the fucking desert. Notable landmarks: none. Recreational activities: drinking alone in the dark. After the three days in Scottsdale, this is where I'm headed: the 10 East to the 387 South. The 387 is where shit gets real: tumbleweeds and cacti and a lonely highway as straight and hot as the devil's dick. As I drive along repeating my mantra, my mind runs away from me.

. . . alone on the highway of my football dreams, I keep my thumb outstretched. Cactus, sand, and rock as far as I can see, cleats over

my shoulder, my teeth are grinding, chewing on a coca leaf. I haven't eaten in days and my eyes are getting heavy. I pack more leaves in my mouth so I can stay awake, stay alert, because I know at any time the right truck will come down this road and spot me, wild-eyed with the madness, and it will pull over and throw open the door for me. I stay awake for this possibility. That's all that I have ever needed. The possibility. Faint recollections of youth flash behind my eyes. I don't know what it means. Is it keeping me alive or killing me? This is the madness creeping in. I shake my head a few times and scream the alphabet backward. Do this enough and truth starts to go backward, too. When my body was young, growing, I stepped with a certain rhythm, a certain bounce that I can picture here on this dusty road. In fact I can feel it. I still have it, and without a soul anywhere near this highway I bounce back and forth across the lanes, simulating the routes I've been running in empty streets since 1984, pushing back the tears with speed. Keep running, keep lifting, keep hitting, keep throwing, keep pushing, keep chasing, keep chasing, keep chasing the life you believe in. I'm kicking up dust on this two-lane highway. I haven't eaten in days, but I'm never hungry anymore. I've got enough money to feed a village forever but not enough to feed me for a day. The money in my account gives me no peace and no appetite my mother tells me it should. She tells me to learn about my money, what it can do for me, what it can do for my future, my family's future. But I've got no interest in money and no time for a family. I've got things to do, mother. I pack my cheek with more leaves and dig into the pavement with the balls of my feet. I think I see something ahead. Yes, I see something. But I've been saying this for weeks now, and every time it ends up being a bush or a rock or a broken-down Chevy or a fucking carcass. Stupid little kid that I am, I get excited every time. I'm back to my old optimistic tricks, circling the imaginary defenders that can't cover me out here in the desert or anywhere for that matter. All I need is a quarterback who understands me. That's all I need. So I keep moving, alone down this two-lane highway, wild-eyed and mad with love and fear and pain and hope. Mad and scared that I was wrong *all along, afraid to stop moving, stop running, because it will all be over. My thumb will go down and I'll start learning about my money and I'll start a family and be very secure and everything will be just fine. But I know when all this happens, when I quit hitchhiking for good, part of me will die. That bounce, that rhythm, the patterns in the street, the wild eyes, the pain, the pain I have learned to cherish, it will all be over. I'm not r—*

Blinker, right on 84, and there's my new home: the Holiday Inn. We will stay here for the duration of the season and bus to Las Vegas for our home games. I needn't stress the regularity with which my mantra is running through my head. Casa Grande feels like rural Mexico. I love Mexico and all, but this is no bueno.

Day one of training camp reveals the many ways that the UFL differs from the NFL. One is equipment. In the NFL, everyone gets everything they need. In Casa Grande—affectionately mispronounced "cassa grand"—we have to fight over the piles of face masks and shoulder pads and thigh pads and knee pads on the floor. Fifty men are rummaging through them simultaneously. The equipment manager is an ornery control freak with a wobbly gait and a lazy eye who initially withholds his services so he can show us all who's boss. Simple questions are met with sharp retorts.

—Where do I get a mouthpiece?

—Oh *now* you want a fucking *mouthpiece*? Shit!

I'm not the only one with a mantra.

There are four tight ends on our team. All of us have NFL experience and we all get along. One of my fellow tight ends, John Madsen, played for the Raiders when I was a Bronco. I knew who he was from playing special teams against him. We studied our opponents and watched so much film that I often felt like I knew the men I was playing against. John and I are similar players, too. We both excel in the passing game and consider ourselves pure receivers. And we both gain very little pleasure from blocking. We also share another fun fact: we are both failed experiments of the Cleveland Browns organization. The nameless face I assumed was watching me try out on the indoor turf of the Browns facility was John. They cut him when they signed me. And here we both are, brothers in misery.

Football is football anywhere. There are tenets of the game that do not change. But the unpleasantness of these tenets can be exacerbated by the conditions that support them. In this regard, football in Casa Grande takes on a new luster. The Holiday Inn supplies us with all of our meals, which aren't horrible, but any meal eaten in the dining area of a desert Holiday Inn, no matter how delicious, is seasoned with the flakes of self-loathing.

Our practice facility is ten miles farther west down the 84, and was hastily constructed ahead of our arrival. In fact, it was still under construction when we arrived. It appears to be of the same blueprint as any nondescript industrial office building, framed on the cheap and ringing with a tinny echo. The facility is to be a multidimensional sports complex with soccer fields, tennis courts, baseball diamonds, and football fields, none of which have been completed except for several football fields with freshly laid sod on the far end of the desert complex, about a half mile's walk from the locker room.

The temperature hovers at about 104 degrees and there is a biblical swarm of mosquitoes that descends on every practice. The smell of bug spray dominates the air. Offensive linemen, firmly rooted in their three-point stances, can be seen frantically trying to scratch the eternal itch on their shin bone just before the ball is snapped. As the days wear on, the size of the swarm grows, alerted by the thirsty desert winds.

Upon returning from practice, it is not uncommon to find a shortage of towels for the showers or a scorpion in your shoe. When returning from meetings and getting ready for practice, it is not uncommon to find the contents of your laundry bag damp and musty. The dryer is perpetually in flux. Yes, life is hard in Casa Grande.

No one has it how they want it. Everyone is pissed off. There is an audible hum underneath it all: what the fuck are we doing?

The answer is Football! Football made us all the toasts of our towns. It got us laid. It gave us status. It made us tough, gave us confidence, and this scar right here, and here, and here. Football! The physical talent we were born with pushed us onto the football field. It was a no-brainer, really. Society funnels people into the industries that their talents serve. And this mosquito farm in the middle of the Arizona desert is the runoff. It is the prolonged, agonizing breakup of a lifelong relationship that cannot and will not ever end well.

Two weeks into training camp and my blocking technique is coming back to me. My old friends, Bump and Bruise, are back, too. And the firecracker pop in my helmet! I'd missed them all. They never asked me any questions. They accepted me. And they gave my mental struggle the physical reminder that pain on the outside was easier to bear. Here. *This* is pain. We understood each other. In the hot, dry heat, my teammates and I act out the tactical know-how of Jim Fassel's offense, sweating bug spray and cracking each other in the face, over and over. One day after the next: all days the same. It's the routine of football in the lives of football men that quiets the demons within. It's the routine that keeps them at bay. And it is the end of the routine that we all fear. That's why we're here.

On one mid-training-camp morning, the same as the one before and the one after, I wait for my turn to run a pass route during one-on-ones. When I get to the front of the line, I whisper my route to the quarterback and get down in my three-point stance.

—Set, hut!

I jab step with my inside foot and shake my defender on the line of scrimmage, pushing up the field with crisply choreographed steps. At the top of the route, I square up my shoulders, give a slight head nod inside, stick my foot hard in the ground and make a clean break toward the corner. The quarterback drops it over my head perfectly as I pull away from the linebacker who is covering me. Touchdown! That's how it's done. Satisfied that my endless pursuit of football perfection has finally been reached, or is finally revealed as unreachable, the hand of fate steadies, lines up the scope, and pulls the trigger. No doubts this time. The sniper hits his mark.

Thwap! My hamstring explodes as I decelerate. I hop twice on my opposite foot, drop the ball to the grass, sit down next to it, and pop off my helmet. A mosquito hovers at eye level. It's over now. It's all over.

Epilogue

The Flood
(2014)

Hi, Nathan. I'm here to prep you for the surgery.

—Huh?

I'm back in Vail, Colorado, lying in a hospital gown a few days after the last down of football I'd ever experience. Not that I know that yet. My hamstring is FUBAR and needs to be surgically reattached to the ass bone. And a PRP shot isn't going to do it this time. They need to cut me open. That's according to good ol' Dr. Philipon. The nurse who has greeted me is holding electric hair clippers.

—I need to shave the area. Roll onto your side.

I thought they'd wait until I was sedated. Guess not. I roll onto my side and she rolls up on a chair. The wheels sing a bitter song on the linoleum. The next tune is a buzzing Norelco. Then an anesthesiologist walks in, swinging a pocket watch in my face.

—You're getting verrrry sleeeepy.

When I wake up my lips are swollen and numb from lying face down, and I have a four-inch scar in my gluteal fold. My mother has flown in to help me during the post-op week in Vail. Then my mom leaves and I'm on my own.

The rehab is in Denver at the Steadman-Hawkins Clinic downtown. My therapist is named Kristen. We're the same age. I think she wants me. Or maybe I just want her. Our sexual tension makes for an enjoyable recovery, but otherwise I'm lost. I'm having trouble talking to people.

On one of my first days of rehab, I'm doing pool exercises next to a man recovering from ACL surgery. He says that the ligament they put in his knee is from a cadaver. I ask him if that makes his wife a necrophiliac. His look of confused disgust reminds me how poorly adjusted I am to civilian discourse.

Eventually I'm cleared by the doctors and say good-bye to Kristen. We both decide not to say the thing we most want to say. Or maybe it's just me. I drive back to San Diego to revive the dream . . . again. People ask me if I'm still training.

—Uh, duh. Yes, of course! Why wouldn't I?

I am still a football player, and I am going back to the NFL because I fixed the problem.

My friend Billy, in whose refrigerator I stored my vials of HGH, has moved from Rancho Santa Fe to a house in La Jolla. He has an open room and takes me in—again. The house is a half block from a wooded area concealing a cliff that plummets to the sea. There's a dirt path along the rock's edge, all the way around the outcropped shoreline of La Jolla. It's an ideal backdrop to exorcise my demons. Problem being that I'm still spoon-feeding them applesauce.

Halfway through the summer, I'm possessed. I tell myself that I feel great. I'm ready to go. What's the problem? My agent, Ryan, says there hasn't been much interest. He says he'll make some more calls. He says someone will bite. He doesn't sound confident.

Around that time, a colony of baby seagulls starts peeping and chirping from a chimney on the roof, beneath the white shadow of their mother. She stands on the chimney and squawks at me every time I walk to or from the house, on my way to the gym or to the local fields where I run in the afternoon.

One morning I open the blinds on the windows to my room, and one of the babies is on the awning above the front door. Little seagulls are mobile at birth, nearly full grown, but dark gray and incapable of flight. Big Baby is stuck.

I go outside and look up on the awning and see the poor thing pacing back and forth. Mama, from above, blinks twice, screams, and swoops down on me with pterodactyl jaws. I turn and run down the curved brick steps and jump in my Denali.

When I return later that day, Mama comes at me again. I duck, pirouette, and sprint through the door, collapsing to the hard wood.

Back in my room, I open the shades to the two windows and watch Baby swaying back and forth in the San Diego sun, helpless. I sit on the bed and hear Mama squawk my name. I look up to see her fly by my window with slow exaggerated flaps, head turned in on me, black eyes burning. I jump up and drop the blinds and turn around, helpless myself. Standing in front of me is an angry seven-foot-tall seagull.

Is this what CTE feels like?

Chronic Traumatic Encephalopathy, or CTE, is a brain disease caused by repetitive head trauma. Football is repetitive head trauma. People always assumed the football helmet protected the brain. Turns out they were wrong. CTE is popping up in the brains of dead football players. CTE makes the walls close in: early onset dementia, depression, memory loss, you name it. No one knew about CTE when I played. Well, that's not true. Some folks knew about it but decided not to tell the men who were affected most. Might fuck up the product on the field. Might make us think twice about smashing skulls. They gave us too much credit, I think. We smash skulls because we like fucking people up. Football players are drawn to the violence. I sure am. I just need to find someone to hit.

In the meantime, Billy agrees that something has to be done about the birds. He says he'll take care of it. It's been a week since Ryan said he'd make some calls to teams. I haven't heard back from him. I call his office and he's not there. I call his cell phone and he doesn't answer. I leave a desperate message. I wonder if he can hear the birds in the background.

The next morning I hear a pecking at the door. I open it. A small, pleasant, dreadlocked man named Nick stands beneath me. He's a freelance wildlife whisperer. Mama introduced herself already, he says, but he wants to have a look at the chimney. I take him upstairs to the balcony and into a Hitchcockian tornado. Mama has alerted the whole clan. She apparently knew it might come to this.

We walk down to the van. While Nick puts on his helmet and gloves, I meet the aura that is his wife and their three daughters: Sun, Star, and Moon. I pick a flower for Star while her daddy climbs onto the roof, pulls a baby out of the chimney, and walks it across the street. He comes back and gets another baby out. That's all of them in the chimney. Then he climbs out of my bedroom window onto the metal awning and, unable to grab Big Baby, pushes her off the ledge. She flaps and lands on the porch. The swarm overhead blacks out the sun. Nick scoops her up and takes her across the street to the woods that hide the cliff, all the while being circled and dive-bombed. Beaks click and clack off the plastic of his helmet. He returns sans bird, and the job is done.

All's well that ends well. I say good-bye to Nick and his universe and look out my window at the birds' new home in the distance. They're going to be just fine out there, I think. They'll probably prefer it, actually. It's much better than a cold brick chimney. They're birds! They want to be around trees and cliffs and dirt and bushes and the ocean! We've all done a good thing, I tell myself.

But then I see a small gray figure waddling out of the woods. Baby is coming home. Mama and kin gather around her, shouting instructions, but Baby's stuck on the corner of a busy street. She walks out a few steps, then gets pushed back by an oncoming car. The sun sinks as Mama keeps guard. Baby paces. I leave the house. I can't watch this.

On my way back later that night, I'm nervous. You can't fuck with nature like this. I ascend La Jolla Village Drive and turn right

on Prospect, preparing for the quick left onto Billy's street. My eyes
are pulled to the corner where Baby was stuck. But Baby is gone.
No, she isn't. She's fifteen feet away in the middle of the street, already stiff: a precocial bloodstain.

I walk through the heavy air to Billy's front door and feel Mama's black eyes on me. I feel the vigil. The guilt. The next morning
I go for a run and avoid the front door altogether, take the side exit.
Mama, who hasn't left the chimney, chases me down the street.
She will never forgive me. I may never forgive myself. I have to get
the fuck out of here. Training camps are in full swing. My phone is
dead quiet. No one wants me. And here I am waddling back across
the street during rush hour. Time to move on or else. Time to face
the end. Time to lay it on the table and smash the motherfucker to
a million pieces.

You have to destroy it or it will destroy you.

One month later I'm in Spain for my buddy Barrick's bachelor party.
He's getting married in France in a week, to his Basque señorita,
Beatriz. We have a few days of partying ahead of us but I'm typing
in the lobby of our hotel in Barcelona like it's my job. I want to write
something about Campdick: the sexual frustration that's created
during training camp by female-deprived NFL athletes.

My friends stop by the cubicle on their way out.

—You coming?

—Naw, you guys go ahead. I'll find you later.

Specifically this little ditty is about the New York Jets' treatment of Inez Sainz, a super-hot reporter who was greeted in the
locker room with snorts and chortles. Then out on the field, errant
balls landed at her feet, over and over again, pure coincidence. I
read about it on the Internet and my fingers got to tapping and it
feels good. I'm purging myself of something. The torture endured
by the writer is often followed by a manic state, unleashed into the
synapses by closing the computer and feeling that the writing was

good. I press "send" and wander off in search of my friends, light on my feet.

I find them at a restaurant. Then we take a cab to la playa de Barceloneta, where the lapping Mediterranean colors a very un-American beach scene: bare breasts and Pakistanis selling cervezas and Senegalese herb dealers hanging out by a giant fifty-foot-tall art installation at the edge of the sand—four big rusty, metal boxes with windows on them. We say hello to our friends from Senegal and Pakistan and plop down on our hotel towels. Asian women yell, "Massaje!" up and down the beach but we do not want a massaje. Soon it starts to rain and everyone runs for cover. We sprint through *la lluvia torrencial* and fly through the first door we come to, which opens into an Irish pub.

We find a nook, sit down, and order beers. There's a group of rowdy Irish girls slamming into each other and yelling. Their leader is a husky brunette with a wide stance and the eye of a tiger. Her name is Christi. She tells me she plays rugby. I tell her she doesn't look all that tough to me.

—Oh, you don't think so, Mr. American footballer? I could take your best shot! I can guarantee it!

—Bartender! I think she's had enough. She's talking gibberish.

—Oh, come now! Don't flatter yourself! Here. I'll pick you up! Come on!

She squats down and motions for me to take a seat on her shoulders. She's sweaty and her hair is sticking in clumps to her neck and shoulders from the musty air.

—Come on! Or do we need to find you a helmet?

Reluctantly, I straddle her neck. She braces herself with a hand on a chair and stands up with a long grunt. The pub cheers. I raise my pint. Christi yells something Irish, loses her balance, and we topple over a chair onto a table. Glass shatters, condiments fly. We stumble to our feet and hug.

Our Senegalese friends are here too, sitting in the corner smiling. They like Americans, one of them says. They invite us outside, and between raindrops we share a spliff and a conversation in our remedial Spanish. Their Spanish is better. One of them tells us about his grow operation in town: four different apartments. Best shit in the city. Hash is fine and all; that's what most Europeans want. But the green bud: that's what he loves the most. They've been here for two years, he says. Came over from Morocco.

It is mid-September. For the last fifteen years of my life, I have been in pads in September. Weddings, birthdays, holidays: everything was trumped by football.

That life kept me insulated. Kept me incubated. So when Barrick invited me to the bachelor party and wedding, I said thanks, but I probably can't go. I was used to saying thanks, but no. Thanks, but football.

And yet here I am; drunk in the streets of Barcelona. No meetings tomorrow. No game plan to go over. No film to watch. Nothing to prepare for except life. Time to use these wings.

Next stop: Ibiza. We arrive late to our hotel after a delayed flight. Two of Barrick's English friends are already in the rooms and point us to the party favors. We shower up and hit the streets. By the time we arrive at Club Space, we're alreading floating down the ruby river.

My feet are three inches off the floor. On the way up a small flight of stairs, a Spanish girl steps in front of me. I'm one step below her. We look at each other. She speaks over the music. I try to answer. We laugh. A golden thread connecting our lips is pulled taut, and we kiss. The music nods in agreement. We break apart and exhale, and she motions to her friend, who walks over. Their golden thread tightens, and they kiss. Then we all kiss each other, back and forth; three tied tongues, three blind mice. We find a rhythm, rolling in and out of each other, biting, taken over by the animal. Then I taste blood in my mouth and pull away. There's blood on her bottom lip. It glows purple in the black light. She licks it. It blooms again. The first girl sees it and grabs her away. Another friend yells at me.

—*Cabron! Que coño hace?!*
—*Que coño le has hecho?!*

Her friends tell me to scram, beat it. I whisper *Lo siento!* over the tops of their heads. She smiles with crimson teeth. She is still in my mouth.

The Campdick piece, which ran on Deadspin.com, was a hit. I write a few more pieces in the coming months, and after the NFL season ends, I wake up one morning in New York with a book deal. I guess that means I have to write a book. Here goes nothing.

NFL media is made up of two categories, generally: journalists and former superstars. The journalists are there to write the stories; the superstars are there to agree. The journalists have no experience in the field, so they often rely on statistics to explain the outcomes of games. The superstars are the personification of statistic-based fame and wisdom. The snake eats its tail. It's a good soap opera. But almost no one who played the game can relate.

I am walking through the streets of New York. I am wondering how I am going to do this. I stop at a cleverly named coffee shop in Union Square called The Coffee Shop. I write a poem about my waitress. She has black hair and sharp bangs, and her triangle earrings dangle above her graceful shoulders and marvelous ass. Her name is Anaïs. We discuss French fries. I tell her I'm writing a book! I'm a writer! I'm only in town for one more day! Does she want to go to dinner?

—Sure. She shrugs.

We meet that night at a crab shack and afterwards walk from neighborhood to neighborhood, bar to bar. New York washes over me, ignites me from somewhere deep. I feel the needle of Fate being threaded, the hand steadying. There is a new world at my feet. But I don't understand it.

Anaïs is a good guide. She loves New York. She believes in New York. She trusts the struggle that it takes to live here. I can dig that.

I want to know everything about everything, and she tells me all she can. We end the night by walking over the Williamsburg Bridge back to DUMBO, Brooklyn, where we sit on a bench overlooking the East River. The lights of Manhattan twist through the shifting currents. I walk her home to her building, where we sit in the lobby throwing toys at each other in the kid's play area. Her roommates are upstairs and she doesn't want to go up there. We say good night, and I take a cab back to Manhattan. The next day I'm in Denver.

I'm a writer now. I pick up the sharpest pen I can find, plunge it into my belly, and twist. Self-purification frees the hand to record the truth in all things. But it must be done in solitude. I sit alone in my house. My Broncos' memories are everywhere. I gather them up and stack them in the basement: helmets, jerseys, footballs, gloves, hats, photos. Out of sight, but not out of mind. I still feel them on my skin. I still smell them. They taunt me mercilessly. The sentences I write bounce off the walls and come back as question marks. I have to get out of here.

I pick up the lease on Barrick and Bea's Venice Beach apartment, throwing myself into the netherworld of memories and loose connections: palm trees and spinning spokes, caffeine and concrete, guitar chords and bare skin. I tape butcher paper to my apartment walls and scribble notes with lines and arrows and drawings. To visit my apartment is to enter the lair of a madman.

Often the only person I speak with all day is the girl at the register of my local cafe who takes my order. I speak so infrequently that my vocal chords atrophy. Dark fantasies play out in my head. I write poems about the women around me at the coffee shop.

She will bring me coffee.
And I will drink it.
And it will be good.
Because it came from her.

I write as much as I can. I wake up every day ready to write. Sometimes that means staring at a blank screen for hours, enticed by a fleeting thought or the fragment of a dream. Sometimes the words flow like a waterfall, and all I have to do is turn on my laptop and I'm sucked into a trance. I look up hours after I've sat down, and there are strangers around, the sun has gone down, there's a new shift of waitresses, and I have no recollection of what I've written.

The more you write, the more you think about mortality. Death becomes the blinking light on shore that can steady the ship. Death is a trusted companion: the moral to every story— Old Yeller.

Or Dave Duerson. Right about the time I got the book deal, the former Chicago Bears great blasted himself in the chest with a shotgun and left a suicide note asking that his brain be examined for evidence of CTE.

We football guys are all tough guys. Pain is nothing. But there's something else rattling around up there in the skull, isn't there? There's something more than the lizard brain. There's a soul. And it is weeping. Commissioner Roger Goodell stands on a trademarked soapbox and juggles a chainsaw, a machete, and an egg: profit, public relations, and the fragile human brain. This is your brain, this your brain on football. Any questions?

If we are to throw aside the war of sensation and semantics and believe the scientists, here's the skinny: Cut open thirty-five dead football players' brains and thirty-four of them display evidence of CTE. That's according to *League of Denial*, the book that chronicles the discovery and cover-up up of CTE in the NFL.

The percentages aren't in my favor. I try to write about the suicides. But I get depressed too. I feel an electric current shooting through my brain when I write the letters C-T-E. I feel brain-damaged when I think about it. I don't want to sit on the porch with a gun in my lap, waiting for my symptoms to appear. I want to live!

• • •

After I arrive in Los Angeles, I file a worker's compensation claim against the Broncos for injuries I sustained while under their employment. Like most former players, I get a lawyer. We subpoena the Broncos to get the medical files that I have used in the text of this book. I answer questions from the Broncos' attorney, who comes to California to poke holes in my case. I visit multiple doctors, submit to multiple scans, X-rays, MRIs, and questionnaires, and read every waiting room magazine cover to cover. I answer everything honestly. It's all fucking documented. My body was destroyed on their watch. But they don't want to pay. They find collective bargaining loopholes in all of the California cases. They push a bill through the California legislature that bars former players from filing in California. They want us all to go away.

The new law requires that 80 percent of your workdays have to take place in California in order to file a worker's compensation case there. I was born in California. I went to school there. I played for the 49ers. I played in Oakland and in San Diego every year for six years. I broke my leg in the Chargers' end zone. I've spent entire off-seasons in California. My agent, to whom I paid 3 percent of my salary every year, lived and worked in California. I live here now. I pay taxes here. But my case, along with four thousand other pending worker's comp cases by players who need medical attention, is thrown out. Now there are disabled former players with no insurance whose care will be funded by taxpayers instead of the NFL owners who oversaw the slaughterhouse that crippled them in the first place. God bless America.

I try to keep my head down and focus on things I can control. I jump on my bike, my primary form of transportation in Venice. The spinning spokes accommodate my spinning mind. Me and my bike! Together we traverse the back alleys and the boardwalk. We cut across traffic. We make illegal turns. We are pulled over by a Venice motorcycle cop. We are deemed confrontational. We are told to step off the bike and put down the backpack. We are asked what we are hiding. We say nothing. We are asked again. We are defiant.

We are told to spread 'em and we are searched. We tense. We want to strike the fool with a lightning bolt. We are handcuffed on the corner of Lincoln and Washington at high noon. We are told that the backpack has to be searched. We laugh and say, yeah right. We are told, fine, we can do this the easy way or the hard way. We say, let's do it the hard way. More cops arrive and surround us. We are fingerprinted on the sly with a digital finger printer. We are cross-checked in the database. We come up clean. We are worn down and agree to a search of the backpack. We watch gloved hands pull everything out and put each item on the curb: a towel, a book, a journal. We watch a love letter blow down the street, and a cop run and step on it with his dirty boot. We watch them find nothing illegal. We laugh. We want blood. We are uncuffed and asked why we were defiant: for the principle? Yeah, for the principle, we say. Ever heard of it?

Fuck the cops. Fuck everything. Just stay moving and all of this will come together. Stop moving and it turns to stone. I crave the movement, the edge of the razor blade. It helps me write. That's why I like coming to New York, to the Ace Hotel in midtown.

> The Ace Hotel gleams and pops.
> The DJ shakes the rafters.
> The waitress glances at the shelf.
> I think of things to ask her.
>
> The mixers mix elixirs quick
> And pick a garnish for them.
> Bar-backs reach a far-back glass
> For cleaner spots to pour them.
>
> The women are black orchids.
> Their hair the sheen that speaks.

Their darkest clothes hide treasure troves.
X marks them, I believe.

A month before I'm supposed to hand in my book, I sit in the
lobby at the long communal table, drinking coffee and writing horny
poetry. I patch together chapters and revise and rewrite. One day
before the deadline, I finish. Some dude at my publishing house
prints out the whole manuscript with a color copy of the cover, hugs
me, and I leave. The extra weight in my backpack is proof of something. I walk down Broadway toward the Ace and feel something
like loss, like sadness, like accomplishment. I have no kids, but I
assume this is like dropping your son off at college and driving away.

That night I have dinner with Anaïs, who has moved from
Brooklyn to Chinatown.

—This place is cool, right? Eastside Radio is right next door. I
live really close.

—Did you walk here?

—No, because I came from Brooklyn, otherwise I'd walk. The
food is really amazing here, Nathan. The spaghetti limone . . . you're
going to die!

—What are you drinking?

—Lambrusco.

—Can I try?

She passes the glass.

—Mmm, good. It's like carbonated wine.

—So . . . I just quit my job two weeks ago: like just fucking
walked out. Someone spat in my fucking face there. I'm sick of being sexually harassed by finance dicks ordering drinks from my tits.
I was celibate for nine months. I was in a weird place last time you
were here. I'm sorry if I was acting like a bitch.

—You weren't acting like a bitch. You just seemed bored.

—No, no, no, two things. Well, no, three things. One, I was on
my period. Two, and don't think I'm crazy, but you were wearing this
lotion, and I don't know why but I couldn't stand the smell. I don't

know what it was. I can already smell that you're not wearing it again, so that's good. Third, and mainly, I don't know, I was just in a weird place. I was subletting my apartment and staying with friends, and I was all fucked-up. You don't know what this place does to you, Nathan. New York City is a hustle. If you take *one* day off you fall behind. Yo, all I need is someone to fund my lifestyle: weed and yoga and fashion and lingerie. I'll be there for him but I don't need him around. I don't need to know where he is all the time. I'll cook for him and wear sexy lingerie, and he can do his thing and I can do mine. We don't have to talk about monogamy. He can do what he wants. I think that's a pretty good arrangement, don't you?

—Yes.

—I'm barely staying above water here, taking odd jobs and stuff. What am I supposed to do? Cocktail waitress, escort, dominatrix, stripper: those are really my only options. I need to make at least one thousand dollars a week to support the lifestyle I want. As of now, I'm not sure how I'm going to pay my rent. I know I will pay it, I just don't know how. There is nothing under the surface here. New York is killing me. I need to get to London for fashion week. I need to, Nathan. I'm just so tired here. I need to get away and recharge. I went to Rhode Island for Easter and stayed with my friend's parents and did relaxing rich-people shit. It was so great. I need more of that. Is that too much to ask?

I return to LA and feel sick. My secret is out. My book, my heart is now in someone else's hands. Please, copy-editor lady, be gentle with it when at first you don't understand and know that I tried. . . .

—Hahahaha!

I hear her shrieking while swinging down the red axe. Back and forth we go all spring and summer. Cut here, snip there. Clarify here, omit there. Don't hang on to your little fuzzy bunnies, your sentimental pudding-pops, your cute turns of phrase. Drown them in the tub. Kill your darlings, they say. Kill them dead.

Writing, it turned out, meant disappearing to the world. Tuning everything out. All forces are opposed to the process of writing a book. Now when I walk into a bookstore, I am overcome with emotion. I imagine how much blood was spilled to fill these aisles, and it hurts me deeply. I have to steady myself on a railing.

But the book launches. A few of the national reviews focus on sex and drugs. Therefore the majority of the radio interviews focus on sex and drugs. Most radio hosts haven't read the book, so they scan a few search results before having me on the show.

—So tell us about the girls! Did they all have vaginas?

—And what about the drugs? Did they get you high?

On January 25, I go back to New York for Super Bowl week. Sell some books. Act like I have job. I check in at the lobby of 30 Rockefeller Center and go upstairs, where I am met at the elevator by an MSNBC producer. We shake hands, and she leads me into a waiting room. I take off my jacket and sit down. A woman comes into the dressing room and looks at me, annoyed.

—Am I sitting in your seat?

—Yes.

—Really?

—Yes, I get ready for my show in here.

—Do you want me to leave?

—Yes.

I have no idea who she is. I leave her dressing room and find another room. Before long I'm summoned to hair and makeup. I'm going on MSNBC with anchor T. J. Holmes in twenty minutes. Those itty-bitty two paragraphs about marijuana in the book have made me something of a "weed specialist" over the last few weeks. I have appeared on *Real Sports with Bryant Gumbel* to talk about my experience with the plant while I was playing football. Current players can't talk about it. And a lot of former players don't feel comfortable talking about it either. I don't mind anymore. I'll talk about anything.

They mic me up and lead me down a hallway, through a door and into another hallway, where two women sit in front of monitors with headphones on. They mouth hello. I smile. We keep going down that hallway and around the corner to the soundstage. Cameras swivel and slide. Men in headphones and horn-rimmed glasses nod. Talking heads spit *politick*, wax relevant, make a case. The segment ends. These talking heads have bodies, it turns out, and feet, which they use to stand up and walk away.

Time for the stoner jock. I take my seat and shake hands with TJ. He thanks me for coming on the show. 5-4-3-2—Live television is terrifying—1. TJ reads the intro on the teleprompter:

When we talk about drugs and professional sports, it's usually us talking about performance enhancers. Smoking marijuana before the game is not considered something that's going to help you perform better. But what about after the game? It could be a huge help. That's according to at least one former player who is raising eyebrows by advocating for the use of marijuana as a viable painkiller option in the NFL.

—You played in the NFL for six seasons, Nate, did you use marijuana the whole time?

—Yeah, pretty much. When I needed it. When my body started to break down as the season wore on.

—How prevalent is marijuana use recreationally in the league?

—It's hard to say. Maybe half.

—And how prevalent is it for medicinal purposes to relieve the pain?

—I'd say that's why the guys are doing it, subconsciously.

—Well, what's wrong with the pain medication that the team provides?

—Well, they're not for everyone. And I think they're bad for you. My experience with pain pills like Vicodin and Percocet— I didn't like them. They weren't good for my body. But they're really easy to get. That's what the doctors give out. And they're highly addictive. A lot of guys get hooked on those and leave the game with a serious addiction. Marijuana was not like that for me. I had no physical withdrawal symptoms from it, and it alleviated the need for pain pills.

—You know what they're gonna say: these guys just wanna get high. They just wanna get *high*!

—Ha, well if getting high relieves pain, then yeah, they want to get high. But isn't that what pain pills do? They get you high in a different way. They don't heal the injury. They just take your mind off of it or dull those pain sensors or whatever. And I think marijuana has a similar effect.

—Now, marijuana is a banned substance in the NFL. Considering that, do you feel that you cheated the game?

—Ha, no, I don't think I cheated the game. Like you said in the intro, it's not a performance enhancer.

—All right, we'll leave it right there. You . . . stopped using after you got out of the league, right?

—Ha, yeah, of course!

—Not using now? The pain is gone?

—Yeah, it's gone!

—Why are you smiling like that?

—Why are *you* smiling like that?

From TJ's show, I go to 56th Street and 7th Avenue and whore myself out on Radio Row, which occupies a major hotel's convention space. Every sports radio show in America descends upon the Super Bowl city for the whole week prior to the game and sets up shop in one big, sweaty room; table to table, ass to ass. Then they look for guests. Many of the producers have seen the *Real Sports* piece. They see me walking around and ask me to come talk pot. I am happy to oblige.

Who can laugh the loudest? Who can tell the most macho story? Who can book the most famous guest? Who can rattle off the relevant stats? Who can be the moral compass? Who can predict the future? I marvel at the conviction with which radio personalities lambaste players. I marvel at how thoroughly they deify the profession.

They all ask me about drugs. I give it to them straight. There's one drug that's more dangerous than all the rest. It's called football. But there is glory in that drug. There is virtue. There is honor in throwing yourself on those train tracks. A lot of people make a lot of money from that sacrifice. In fact every motherfucker in this cavernous, hangar-sized hotel ballroom makes money off of that sacrifice. They depend on it. They need it. And they will manipulate every argument to ensure that the virtues of football remain unsullied, so that their jobs and their names stay unsullied too.

The Radio Row circle jerk pulls the life out of me. So does the rest of the city. My voice is gone. My brain is mush. My lips are chapped. New York wins again.

I fly back to California, and I watch the game with my family on an old school-projection television in a rented beach house near Santa Cruz. It is my dad's eighty-fifth birthday. I eat pizza while the Broncos get rolled by the Seahawks. As the camera zooms in on a perplexed Peyton Manning, my brother asks me if it's sucky to see them lose like this. Not really, I say. I mean, it sucks for them and for the fans.

—I want to see them do well and all, but it doesn't affect my mood. They're playing a football game. One team will win and one won't.

I sip a beer and look out on the ocean.

After the Super Bowl I drive back to LA on Highway 1 through Big Sur. I stop at the Henry Miller Library. Henry Miller held a dagger in both hands. Henry Miller slashed and killed. It takes balls/ovaries to write what you want, exactly how you want, in spite of the world, in spite of relationships and judgments and expectations. The Library is located at one of the thousands of bends in the coastal road. The grounds used to be the house of Miller's friend Emil White. "The Henry Miller Library: Where Nothing Happens." It has a large open yard with a stage on one end. I step out of the car and into the mud. The redwood trees shoot straight up all around me; bug ladders to god, casting soft blue shadows over the swaying ferns.

Smoke hangs over the yard. I walk down the winding dirt path, lined by rocks and occasional sculptures. At the base of a redwood is a twelve-foot-tall crucifix. Through the trees, the sun sweeps in and slices my vision into smoky green pieces of rhubarb pie. I squint through the pulp at the crucifix. It's made of old computers: Apple 2Es. I take out my iPhone 5 to snap a picture, squat down. Just when I've found my frame, I feel a nudge on my butt. I look down and a black-and-gray cat is inspecting me.

A man and a young woman run the shop. The young woman volunteers. They scurry about outside with the cat, tending to the communal fire, and setting up chairs while I peruse the small store. It smells like burning wood and old paper and seawater and dirt. Being here makes me want to lock myself in a cabin and write. So much blood in every single book: a man's entire life. A man's sense of worth, his agony and tension played out in each word of each page of every book from both ends of eternity inward. It's an ominous thought: that any book I write will be just a grain of sand in this desert. But I love this desert. I live here now.

I buy a few books and talk with the girl. She moved to Santa Cruz from Michigan and loves this coast, this area. She possesses the peace of mind that I've always envied: knowing who you are and where you want to be and not longing to be anywhere else.

I drive on. Over the next three days, in LA, I record the audio book of *Slow Getting Up*. "It's over now. It's all over." It feels good to read these words aloud. That night we have a soccer game under the lights in Santa Monica. I'm getting back in touch with my body and learning how to move again in a more fluid fashion. Football trained me to be a linear exploding robot. I lost my ability to be supple and patient. I lost my natural athleticism, my finesse. They are coming back, slowly. But I also still have a psycho in me. Sometimes I have tunnel vision and attack the ball and run over someone. I feel bad when that happens. But I also feel good. I write a poem about the feeling good part.

It feels good to hurt you.
It feels good to cut you down.
It feels good to watch you bleed.
I'm thinking of it now
And it makes me smile.

(Am I a sick man now that I've said it?
No. Words are nothing.)

It feels good to watch you cry.
It feels good to crush you.
Your pain gives me so much pleasure.
I wish I could quantify it.

Anyway. The next night I go to play music with my friends. We haven't jammed together in a while. We drink magic mushroom tea and plug in. God presses "play" and the music flows in from all corners of the room, rippling the air with purple chords and a bassline sent from the future. We do not have to try. We pluck the plasma and laugh ourselves into a sonic flower that folds us in origami and sets us adrift in the mushroom sea.

And I mean adrift. As we stand outside in the fresh air, laughing at the tree, my phone buzzes in my pocket. I squint at the wiggly lines and the white light. My mind adjusts:

—Nate, this is your neighbor in Denver. There's an emergency at your house. I'm calling you right now.

I weigh the words in silence. My phone rings.

—Hi, Nate. I am standing outside your house, and your garage door is open and water is pouring out into the street. Another neighbor saw it and she told me and . . .

I feel blood running from my ears.

—What should I do? Should I call 911?

—Uh! . . . Yes! . . . Do that! We have to stop the water!

—Okay, I will and I'll call you back.

She calls the fire department and I call Matt Mauck, who threw me my first preseason touchdown of my career ten years ago and is now my dentist. He lives around the corner and arrives at the same time as the firemen. They pop the lock and go into the basement and shut off the water. I speak with the fireman on the phone, and he gives me the skinny: busted pipes; *lluvia torrencial*; flood; lotta water. I manage to communicate well enough to put everything off until the day, when I get a flight to the dirty swamp. Charlie Adams picks me up, and we go together to assess the damage.

As we pull into the driveway, I'm not breathing. There are thick beads of condensation on all of the windows. I turn the key and enter the flooded grounds. The night before, there were five inches of water in the basement and two inches on the ground floor and water raining down from everywhere. That's what Matt told me. But now there's no more standing water on the first floor. It's all been absorbed in the flooring and dry wall and has made its way down to the basement, through the floor vents, and drowned my hoarded memories before exiting through a small drain next to the busted furnace—the culprit.

The pipes burst in seventeen places when the water in them froze. It was minus-ten degrees in Denver a few days ago. When the weather warmed up, the frozen pipes thawed, and the water flowed freely through the seventeen holes. It was thirty-six hours of indoor rain, enough to collapse several sections of the ceiling and short circuit the garage door, which thankfully opened and alerted the neighbors.

I walk through the house, and water squirts up around every step. A mitigation team comes and starts the dry-out process: big reverse-humidifier thingies. Everything below the second floor has been ripped out, down to the two by fours. There are no walls, no ceilings, and no floor.

All of my Broncos' artifacts were in the basement: jerseys, helmets, clothes, pictures, letters, everything. Not sure what I was going to do with them before the flood. That part of my mind was sealed off. They would have sat down there until I sold the house. Then when I moved, they would have been put in storage. Then when I died, they'd be someone else's burden.

As I write this chapter, I am sitting in my hotel room at the Towneplace Suites, room 325, paid for by my insurance company. It is an extended-stay hotel behind an enormous Ikea in a business park south of Denver. I can see Park Meadows Mall from my window. My TV has HBO and swivels. I hear two neighbors having sex on my way down the hall after dinner. She says, "Don't stop fucking me!" Then it gets quiet. He stopped fucking her.

The carpet in my room is dark red with yellow lines connected by yellow circles, parallel strings of dirty pearls. The pearls have smaller yellow circles in them. They have breakfast downstairs every morning from seven to ten: cereal and green apples and eggs out of a package.

My Denali is in the shop. Its transmission went out. It was having trouble getting out of first gear. Turns out there was no more second gear. I don't know when it'll be ready. It's a one-mile walk to anywhere. And really, I don't mind. It gives me time to think; to pore through every detail of my life, dust it off, place it on the table, and smash the fucker to pieces.

Acknowledgments

This book was brought to life by the support of many wonderful people. The athlete's body is coveted. His mind is implored to stay silent. But the athletic mind is an abundant source of artistic revelation. These people encouraged me to tap it:

Mrs. Namba, my teacher in the third, fourth, and fifth grades at Grant Elementary School in San Jose, laid a foundation of compassion and confident expression. Those remain my most important school years—the most lasting and most complete.

At Bret Harte Middle School and Pioneer High School I was a social jock. Academia was a thing to be endured between practices and parties. But those years made a lasting impression on my heart, and everyone involved lives in the spirit of this book—girlfriends, friends, classmates, coaches, teachers, and parents.

The summer before my sophomore year at Cal Poly, a childhood friend committed suicide. In the days that followed, my mother gave me a journal. I asked what I was supposed to do with it. She told me to write, it didn't matter what—just write. And the words started flowing. Uncapping the pen uncorked my heart.

The next year I transferred to Menlo College and enrolled in a newspaper class, and professor DeAnna DeRosa soon gave me a column in the *Menlo Oak*. There were no parameters on the content or style of my articles. Menlo gave me artistic and athletic freedom that allowed me to flourish.

When the Broncos sent me to NFL Europe, I was asked to keep an online journal for their website. Again, no restrictions on content or style. I wrote for the website for the next three years. I am thankful to Pat Bowlen and the whole Broncos family for allowing my self-expression.

During training camp of 2006, a writer named Stefan Fatsis was given unrestricted access to the team in order to write a book about life in the NFL called *A Few Seconds of Panic*. We fell in together, a writer to a writer—bouncing ideas off one another. In the years that followed, Stefan critiqued and promoted my work, made calls on my behalf, gave me advice, and motivated me to keep writing. He pushed me through the door and into the light.

During two consecutive off-seasons after meeting Stefan, I enrolled in writing classes at Denver University. My professors and classmates encouraged me to believe in the voice in my head and shed my football armor, for which I was not ready, but I thank them for trying.

I owe a special thanks to Tommy Craggs of *Deadspin* and Josh Levin of *Slate*, who, after my NFL career ended, gave me a forum to write about what I knew. Soon I was on a plane to New York to pitch my book idea in meetings arranged by my new agent, Alice Martell. Alice found me by chance, and what a lucky man I am for it. Her compassion and thoughtfulness has made the transition from athlete to writer as smooth as possible, and has given me a big picture perspective that I desperately needed.

That trip earned me a book deal with HarperCollins and an introduction to my editor, David Hirshey, and his associate editor, Barry Harbaugh. We shook hands and became partners. But what started as business has evolved into friendship. This book is a product of that evolution. David and Barry let me find my voice without telling me where to find it: a gift I will keep forever.

After a failed attempt at writing in Denver, I packed up and left for L.A., where I wrote this book in steadfast seclusion. Friends, family, lovers: I turned away from everyone to focus on my work. To all of those people, thank you for your patience and understanding.

Many thanks are also due to the west side establishments that provided me food and coffee and ignored my brooding presence: The Cow's End, GTA, Abbot's Habit, Intelligentsia, 212 Pier, 18th St. Coffee House, and every public library in West Los Angeles, including the always entertaining main branch downtown. The library dwellers and Venice street kids provided me endless inspiration to complete this book.

I moved in a daily loop on my beach cruiser from Washington Boulevard in Marina Del Rey through Abbot Kinney in Venice and up Main Street in Santa Monica, back up to Lincoln Boulevard and south toward home, where I checked my mailbox for brainfood from my pen pal, Vanessa. Her steady, vulnerable honesty allowed me to be honest with myself, a gift for which I can never repay her.

At night I went to my second family's home for food and more counsel. Barrick Prince and Bea Poirier fed me and listened to the daily ramblings of a madman. A few times a week I plugged in a guitar and jammed with Colin Kelly and Ged Bauer. The spirit of the jam lives in this book.

Some days I was elated with a breakthrough. Others I thought I was worthless, doing nothing, hopeless, lazy. Then one day, I looked up and I'd written a book. I took a deep breath and turned back to those I'd turned from, and they were still there for me.

And lastly, thank you to my eternally supportive and loving parents, Ross and Marilyn; to my brother, Tom; and to the rest of my family and friends. From childhood to manhood, the love has been constant, and has shaped my outlook on life. I was a lucky boy. I am a lucky man.

Writing this book was like pulling a huge splinter out of my body. Thank you to everyone who helped me build the tweezers.

About the Author

Nate Jackson played six seasons in the National Football League as a wide receiver and a tight end. His writing has appeared in *Deadspin*, *Slate*, the *Daily Beast*, *Buzzfeed*, the *Wall Street Journal*, and the *New York Times*. A native of San Jose, California, he now lives in Los Angeles. This is his first book.